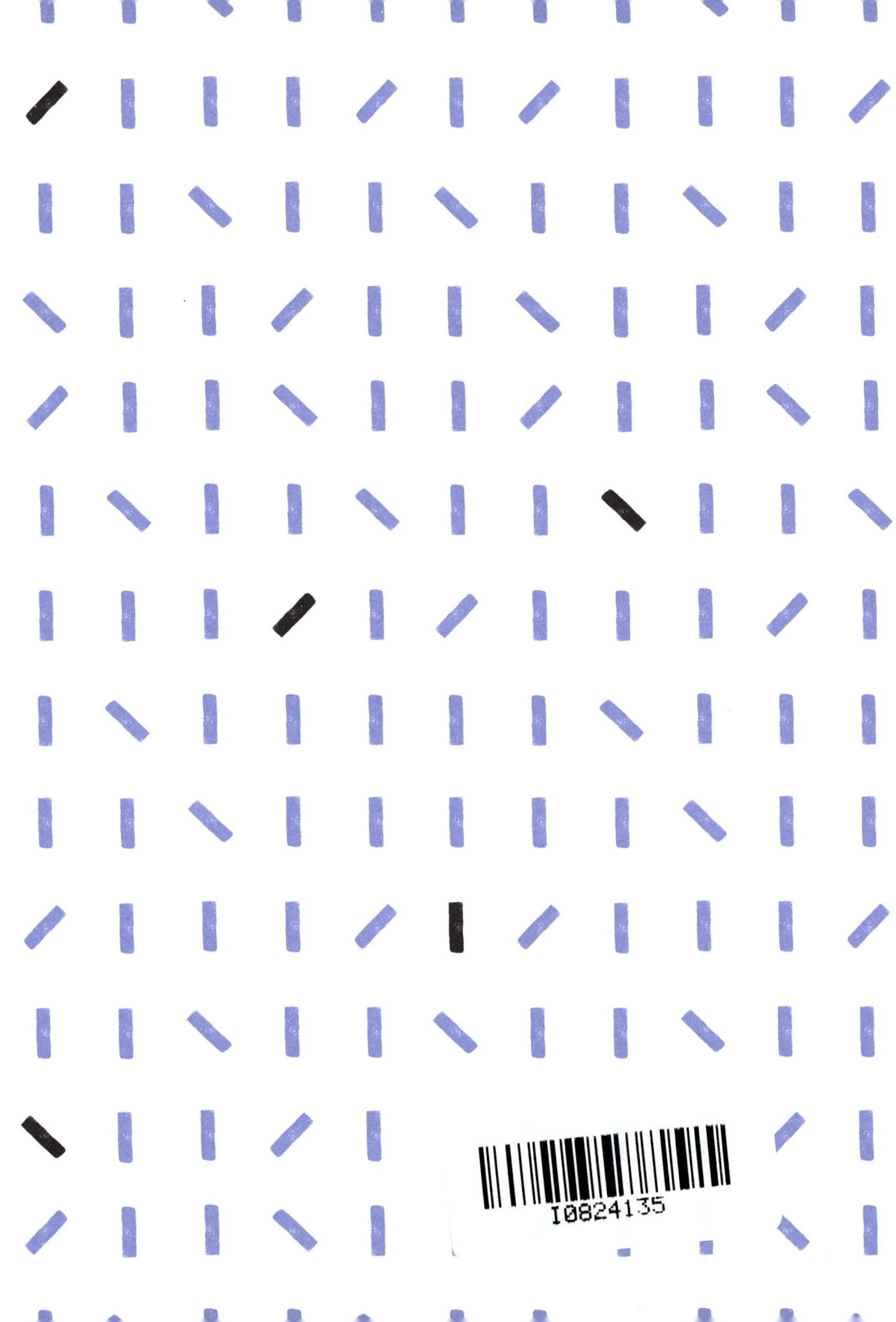
I0824135

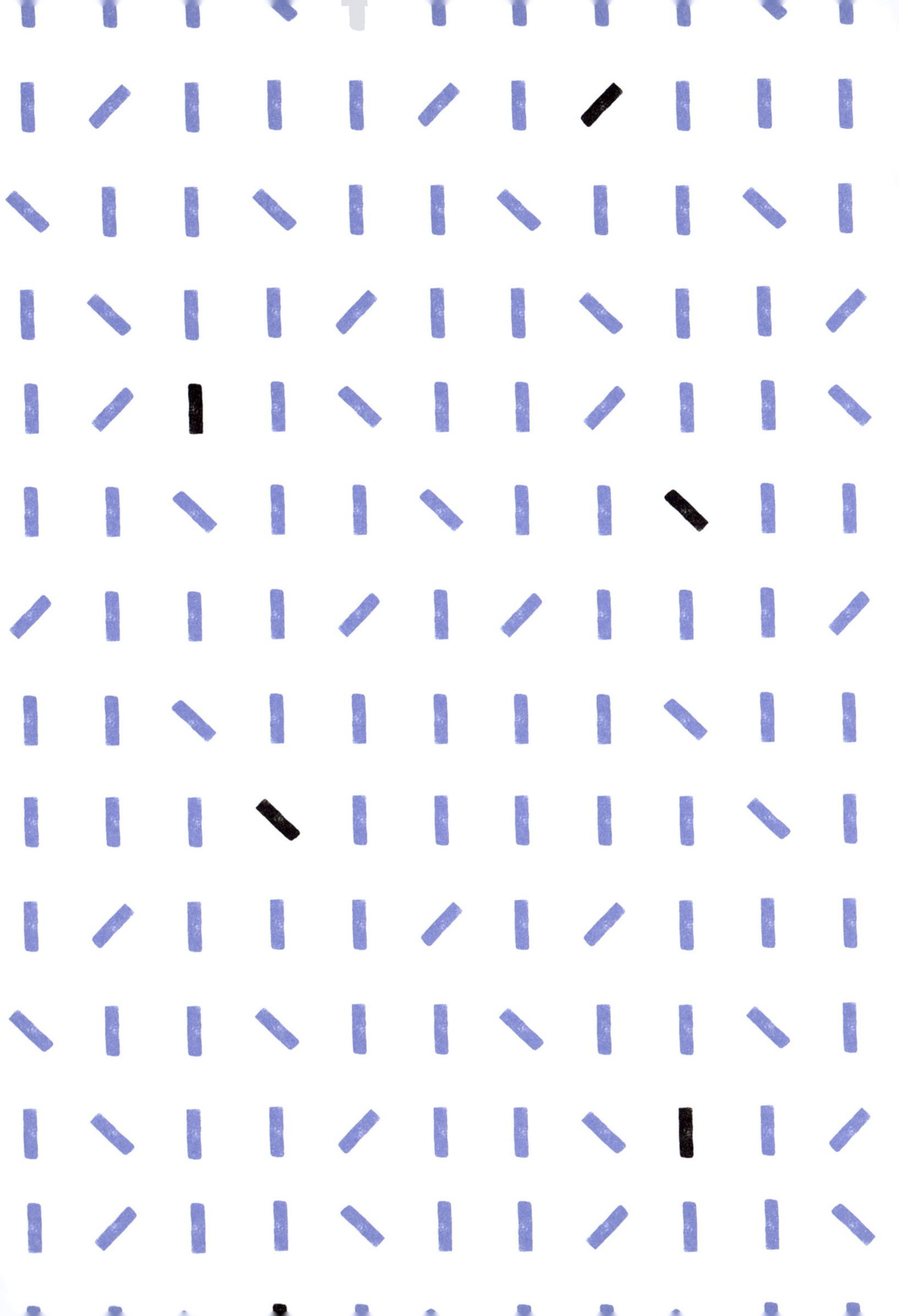

Healing Ourselves and
the Planet We Call Home

KATHARINE K. WILKINSON

The authorised representative in the EEA is Simon and Schuster Netherlands BV, Herculesplein 96 3584 AA Utrecht, Netherlands. (info@simonandschuster.nl)

Amber Lotus
an imprint of Andrews McMeel Publishing
a division of Andrews McMeel Universal
1130 Walnut Street, Kansas City, Missouri 64106

www.amberlotus.com

26 27 28 29 30 GPS 10 9 8 7 6 5 4 3 2 1

ISBN: 978-1-5248-9989-9

Library of Congress Control Number: 2025949147

Editor: Melissa R. Zahorsky
Art Director: Julie Barnes
Production Editor: Kelsey Rolofson
Production Manager: Beth Steiner
Book design by Ampersand with special thanks to Lydia Sweeney

For the persistent shimmer of life

And the world cannot be discovered by a journey of miles, no matter how long, but only by a spiritual journey, a journey of one inch, very arduous and humbling and joyful, by which we arrive at the ground at our own feet, and learn to be at home.

—Wendell Berry, *The Unforeseen Wilderness*

Table of Contents

Pre-Ambling

Welcome to Climate Wayfinding.
I'm glad you're here.

It is no small thing to be human on Earth.

If you have found your way to this book, I suspect it is because you, like me, feel a deep ache about what is happening within our web of life. Perhaps you glimpse news of the latest unnatural disaster and look away to remain afloat. Or you survived one and wonder how to rebuild. Perhaps you boil with outrage at those who gamble our lives. Or you boiled so long that you've run completely dry. Maybe you want to grab a friend's hand and say, *Do you see how bad this is?*

On some level, most of us wish that nagging ache would go away. Its rumble complicates our already complicated lives. *This is too much,* we whisper. And that is true: It is entirely too much.

But if we feel into those feelings, we can also touch a desire to make things better. It reveals that we carry within us the impulse of life. We want to care for the people and places we love. We want to arc toward healing—healing ourselves, our communities, and our breathtaking, singular Earth.

We often struggle, though, with what path to take.

Over the past twenty-seven years and counting, I have been on my own winding journey as what you might call a *climate person.* Along the way, I have often found myself wishing for maps that did not exist. I have hit many crossroads and a few outright dead ends. And I have needed support with how to be someone who loves this planet and craves a sense of purpose. At times, the road has rattled me, but it has also been rich with revelation.

I have grappled with key questions while circuiting through studies and jobs and a veritable swarm of climate undertakings:

+ Is there a way to transform my overwhelm, grief, and outrage into power, joy, and meaning?
+ Is there a way to upend my loneliness by forging connection in its place?
+ Is there a way to turn my doubt about what I have to offer into conviction and contribution?
+ And, if it is possible to shift these things in myself, might there be a way to support others to do the same?

Yes, I've discovered over time—albeit in fits and starts.

A few years ago, I began to imagine and design a program that would experiment with these transformations and give them some structure. And then I began building it alongside colleagues at the nonprofit I lead, The All We Can Save Project. We called the experience *Climate Wayfinding.*

Facilitated by our team, groups of a couple dozen people would go headlong into the questions and work toward answers together. We quickly found that *Climate Wayfinding* was offering something potent—and that the hunger for this unique, heart-forward approach was much bigger than we could fill. Even when we trained facilitators across the United States and Canada to bring it to their respective communities, we were barely beginning to meet the need.

That is where this book comes in. The sessions we run in our programs form the basis of the chapters that follow. Each one equips you to take a step in the journey of *Climate Wayfinding*, either alone or together—at home, at work, in the classroom, or in community. (See "How to Navigate This Book" and "How to Gather as a Group.") Most books talk to you; these pages hope to walk with you.

Wherever you are on your climate journey—whether a longtimer or newly curious—the approach of and practices in this book will serve you well. Perhaps you see climate as all-encompassing and paramount, perhaps you consider it one piece in humanity's jigsaw puzzle of challenges, or perhaps you notice it seeping, increasingly, into your life and work. The need to navigate is the same, as are the means to do so.

As a facilitator and teacher, I have witnessed hundreds of people move through this experience and emerge with a newfound sense of clarity, courage, connection, and contribution. I have seen the metamorphosis of paralyzing ache into action. *Climate Wayfinding* offers passage to firmer and more fertile ground—ground where we can be with the wounds *and* be a part of mending them.

Our moments of seeking are uniquely precious in the human experience. They rouse. They goad. They can send us down the slipway of healing and transformation, first in ourselves and then in expanding circles. To find our truest ways forward is, at heart, an act of generosity.

There's a fresh scent in the wind. The terrain ahead is calling. Let's begin.

KKW

How to Navigate This Book

Climate Wayfinding is a participatory experience. To evoke poet Anis Mojgani, "Come closer. / Come into this."

This book is designed to engage the whole of ourselves—*head*, *heart*, and *hands*—and you will find some key elements in each chapter.

HEAD: *Orienting essays* shepherd you through key ideas and reveal where we are going. Key *frameworks* bring structure and synthesis to the tracks of exploration. *"Lighting the Way" stories* share the wisdom of other climate leaders to illuminate the path.

HEART: Curated *grounding poems* and *playlists* meet you in a heartspace, right where you are, and do what art does best: make us more porous. Along with *illustrations* throughout the book, these elements help us access greater wonder, attunement, and tenderness—essential to finding our way.

HANDS: *Prompts* offer generous questions to mull and draw your pen to paper. You'll find them as light "stepping stones" within the essays, followed by in-depth invitations for journaling at the end of each chapter. Freewriting is an essential part of *Climate Wayfinding.* You'll also find *exercises* to spark further self-discovery and small steps beyond the page. Engaging with all of it will take about an hour per chapter. A journal or notebook will be key—for freewriting, mapping, and jotting down insights.

To gain even greater richness in the experience, *group agendas* provide simple guides to gather and engage with the book in community. (See "How to Gather as a Group.") If you are reading solo, feel free to flow right past this section.

Finally, *digital content* enriches the journey with audio guided meditations, interactive tools, additional resources, and more. Whenever you see this compass stamp, know that there is something waiting for you beyond the page at climatewayfinding.earth.

It's possible that some elements of this book will be just your cup of tea, and others might not be. If something doesn't serve you, feel warmly welcome to adapt it or just cruise right along. But/and/also: If you feel yourself a bit hesitant somewhere, see what happens if you aerate that with curiosity and openness. Sometimes our deepest shifts come where we least expect them.

Most importantly, please take care of yourself, dear reader. If you are struggling with anxiety, depression, PTSD, or suicidal ideation, seek professional support. In this experience, go to the depths that feel right and manageable for you, given where you are and the resources you have access to.

How to Gather as a Group

Every cohort I have guided through the live experience of *Climate Wayfinding* has been a reminder of how very capable we humans are of giving each other what we need in this time.

We can offer one another deep listening, presence, compassion, encouragement, and solidarity. We can receive generosity of spirit, new perspectives, and infectious joy. So, how do you take this journey as a group?

First, you'll want to form one. A group can be as small as three people or as large as thirty. (That's about max for being in a single conversation.) Friends, neighbors, family members, coworkers, students, fellow members of shared space, wider networks of folks—all are potential group members.

You'll want to plan to gather eight times in total, once for each chapter, and give yourselves 75–90 minutes together. You can circle up in person or online—either will work well. Everyone will need a journal and pen. Ideally, you'll have a way to play music.

Before each gathering, everyone will engage with the content of a given chapter on their own. Then, you'll be ready for a rich collective experience. The agendas at the end of each chapter will be your guide. They draw on and build upon the reading, journaling, and exploration done solo. Key elements will show up again and again, so they should grow increasingly intuitive each time you come together.

It's helpful to have one or two people shepherd the group's collective experience. If you are playing that role, check out "Guidelines for Gathering as a Group" at the back of the book. Then, be sure to read through the group agenda in advance of each gathering.

Above all, enjoy being on the journey together.

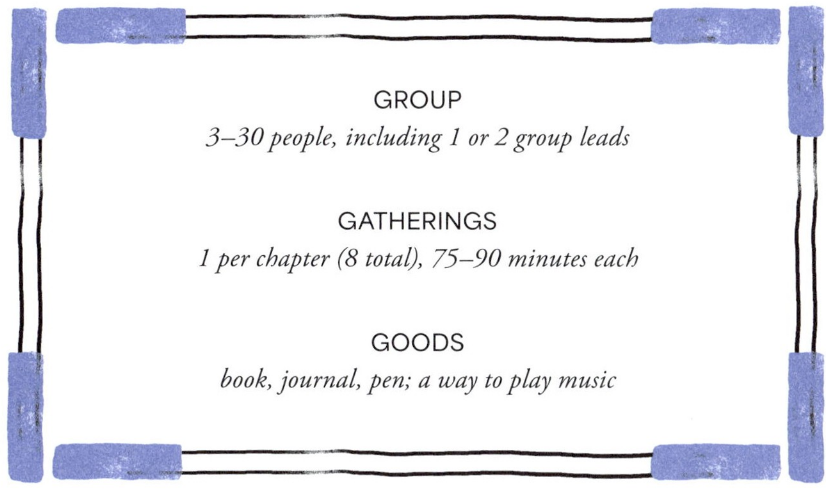

Find additional resources for starting a group at climatewayfinding.earth.

1

Embarking

Come to the crossroads
where your wonderings meet.

At the age of sixteen, perched on a ridge in Pisgah Forest, North Carolina, I scrawled these words into a handbound journal: *Want to help the world. Be connected with the Earth. Change the way I live.*

I remember the day was cold—oaks, hickories, and maples long ago having given up their leaves for winter silhouettes—but the sky was bright. By the time we made it up the valley, my lungs were heaving in the sharp air. A dozen high school sophomores had followed our teacher, Susan Tinsley Daily, through a beautiful, remote stretch of the forest.

We were used to going to magical places with her: crawling through a twisting thicket of rhododendrons and writing about what it might be like to live beneath them; sliding on our bellies through a flooded riverside field, smearing our faces with mud, and cackling with delight as we made face-prints on paper. Her classroom, of four walls or none, was typically a place for growing our sense of aliveness.

But that day, Tinsley led us to a totally denuded ridgeside. The clear-cut was a barren gash in the Southern Appalachian ecosystem I had come to love. I felt my heart fissure as I looked out on what was traditionally—rightfully—Cherokee land, once canopied by towering chestnut trees, before the days of colonization and forced removals, then ubiquitous logging and deadly chestnut blight. So many layers of tragedy.*

The brutality of the extractive economy gripped me. I ached with realization about the gap, the gulf, between how we *can* do things—like accessing timber selectively, with attention to ecological integrity—and how we *do* do things—taking it all, land be damned. I shuddered with shame from not having been more awake before, thinking clear-cuts were a wrenching, awful problem in faraway places like the Amazon, but not here.

This was the most pivotal instance in a series of experiences that had helped me understand the harms of industrial forestry and agriculture, fossil fuels, and climate change. And through that understanding, I also came to see my own entanglement in those systems—unintended and largely out of my control, yes, but painfully present. The mundane use of toilet paper or a plastic shampoo bottle collided with the existential. I was heartbroken, angry, activated, and I yearned for a sense of belonging on, and with, this planet we call home.

My mother has always called the Appalachians "wise old mountains," not as tall or dramatic as their younger relatives in the American West, but sage and

*Roughly thirty miles northwest of where we encountered the clear-cut is the Qualla Boundary, the 56,600-acre territory of the Eastern Band of Cherokee Indians. It is a sovereign nation of some fifteen thousand members, who continue to steward these ancestral homelands and the culture woven into them.

powerful in their longevity. In the presence of these remnants of geologic uplift, now carved by cold water and swathed with moss, one feels summoned to deeper truths.

I recall a steadiness of hand as I recorded my clarity and commitment in that pale-blue journal. My lucidity, cast into vows, has stayed with me for almost three decades. But like many moments of inward truth-telling, the one inked that day prompted more searching than it supplied answers.*

What can I do to help?
What does this mean for the direction of my life?
Will action ease the anguish I feel? Will any of it matter?

That is the paradox of a calling: certainty and uncertainty in the same breath.

What I did not realize at the time was that I was stepping into a lifelong process—a navigational process at the nexus of my own being and the broader world. That kind of navigation is nothing new. Among other core attributes of our species, human beings are seekers and makers of meaning. Our own lives—the time we have in these fleeting bodies and the paths we take—are raw material for that creative act. The terms *purpose*, *vocation*, and *calling* all point to an enduring desire to carve significance out of, or into, the years we spend embodied on this Earth. We want to make something of a heart that pumps, neurons that dispatch, and lungs that inspire—each piece of us, an improbable ancestral gift.

And so we wonder:

What can I offer to our world?
Is my soul built for certain contributions?
Who am I here to become?

* During the hunkered days of 2020, I reflected on this formative experience in "morning pages," one of Julia Cameron's two essential practices from *The Artist's Way*. That free-flow, longhand writing evolved into a piece for *Time* in 2021, some of which is echoed here.

Questions like these, and the longing that can churn beneath them, are a through line of late-night journaling, conversations between best friends, self-help summits, therapists' couches, elders' guidance, and every medium of storytelling. There is good reason many know by heart Mary Oliver's study of the essential question: "Tell me, what is it you plan to do / with your one wild and precious life?"

> Have you held this kind of tender, pulsing question? What shape did it take?

Our Changing Context

Today, the perennial act of navigation has a new context: the climate crisis and all its implications for life, both human and beyond. Those implications are already part of our days: ferocious wildfires and stifling smoke, floodwaters and storms of freakish speed, dallying domes of blistering heat, and likely the warmest temperatures in at least one hundred thousand years. (Back then, *Homo sapiens* were just beginning to use symbols, the first marks made with meaning.)

What scientists once outlined as *someday* climate possibilities are, more and more, our shared reality. Shorelines and thoroughfares slip beneath rising seas. Whole forests and centuries-old towns are reduced to ash in mere hours. Land that sustained specific crops for generations lurches from ideal conditions to inhospitable, snatching the livelihoods of farmers and communities. And entire species—snuffed out.

All of it is happening in a profoundly unequal world, with those who are already in jeopardy or facing injustice, and who have scarcely contributed to this crisis, suffering most. Our new troubles inflame existing ones.

In a very literal sense, we find ourselves facing a world where maps increasingly no longer work.

Our physical world is changing with such rapidity that efforts to document it are almost always out of date. From road maps to planting schedules in the *Farmers' Almanac* to passed-down knowledge, the records we have used to plan, steer, and respond to conditions—they all feel less and less reliable. Have we logged the latest mudslide, algal bloom, oil spill, or refinery discharge? Is it safe to take that road, swim that lake, fish that bay, or breathe that air? Is that grassland or glacier still there? What are the edges of water, of land, of high temps and low, of *safe* and *sure*?

The gap between what is on paper or screen and what we encounter in three dimensions can be startling, distressing, and also dangerous. We are feeling our way through the obscure, a word that comes from the Latin for "dark."

As we traverse, our internal maps are failing us too. *Where to live, what work to do, who to love, what matters most*—these queries press amid the commotion of change that is global in scope yet intimately local in consequence. They seem to outstrip many of our inherited answers, and efforts to track our own deepest knowing may come up murky and muddled.

With *home* so very much in question—as flux comes to neighborhoods, ecosystems, and entire countries—how can we not lose our bearings? In fact, if we feel we have clear coordinates on life, we may not have truly reckoned with what's afoot and the material conditions that will shape every path from here.

For meaning-seeking beings, this time of climate crisis and ecological unraveling is clarifying. If we are looking for something worthy of our waking hours, surely the work of mending our planet is it. The continuance of life itself seems a rather consequential mission—personally, communally, and for society as a whole. But in other ways, this moment is wholly confounding. To be so small, in the face of something so big, is disorienting.

> Take a moment to inhale . . . and exhale. Inhale . . . and exhale. Let breath be a tether to here, to now.

More and more of us are waking up to the climate and ecological crisis now fully on our doorstep. As we do, we find ourselves unsettled, unnerved, and even unmoored. We do not want to be in this mess—yet cannot piece together how we get out of it.

If you think of the climate crisis and feel bewildered, overwhelmed, useless, or inconsolable, you are not alone. Globally, 84 percent of young people are moderately to extremely worried about climate change. They express feeling sad, afraid, angry, powerless, and betrayed by leaders. A majority of Americans are worried, too, and a quarter of the country expresses alarm about our heating planet. But a much smaller group—just 8 percent of us—has turned alarm into action through activism, donations, consumer choices, or even just discussing the topic with friends and family. More Americans say they are willing, but unsure how best to engage. Around the world, 89 percent of people want to see more climate action from their governments—but often do not register that many, many others share the same desire.

The climate sidelines are filled with concerned, but largely quiet, spectators. (Getting to this point was no small feat, to be sure; ask any longtimer in the climate movement.)* Now, we need many more people to join the pilgrimage of planetary healing, which is being routed in real time. And we need to ensure that those who do join can stay engaged over the long haul—which is, I'm afraid, almost certainly what lies ahead. As social justice strategist Makani Themba has said, "We are living in high-stakes times. There is no one to spare. There is no bystanding."

*In the 1800s, scientists like Eunice Newton Foote pointed to the possibility of what we now call climate change. Research intensified in the twentieth century, including Dr. Charles Keeling's pioneering documentation of carbon dioxide in the atmosphere. The 1980s saw a surge in researchers, activists, and Indigenous communities sounding the alarm. In 1988, NASA scientist Dr. James Hansen delivered a seminal warning to the US Senate, and the Intergovernmental Panel on Climate Change was formed. The public concern evident today grew out of decades of work.

Nodes of Possibility

As a climate leader, the core belief that inspires my work is this: *Each of us is a node of possibility for healing the climate crisis—whoever we are and whatever we've got to give.*

In the world of plants, nodes are the points on a stem where leaves or buds emerge. They are the sites of new growth. Nodes are also the points of intersection in an underground mycelial network, where the fungal web grows stronger for its crossings and connections. For flora and fungi, nodes are where the action is. I believe we also have the capacity for that kind of generative potency.

Maybe your fingers are snapping because you believe that too. Or maybe the idea that we can be a planet-healing force feels entirely far-fetched from your particular lookout. Certainly, there is plenty of the antithesis on display. But believer or not, the work remains the same: to allow and activate the forward motion of that inherent possibility we contain.

Just as surely as I'm convinced we can each be a generative node, I also know we rarely, if ever, become one alone. Solo, we may be paralyzed by distressing climate emotions or mired in our own doubt; cheat our immense capacities to help or lose sight of what could be; defer to the current course of things or burn out trying to reroute it. We may simply lose ourselves in our lostness.

What's more, solo action is rarely the pathway to social change. A node, by its nature, is part of something bigger. Cultural celebration of lone heroes—think of Mahatma Gandhi, Dr. Martin Luther King Jr., or Greta Thunberg—is a persistent predilection that misleads us. Lionization of individuals, however visionary and bold, veils the collaboration and collectives that are the fuller, truer story. That fuller, truer story is one we need now. We need a story of *our* possibility, a story of *us*.

> Pause on the word *possibility*. Welcome in the maybe-ness of it. Can you feel possibility in your body—perhaps a warmth, a quiver, or something else entirely?

When we scan the current landscape of society's leaders—in a conventional, narrow definition—we see climate shortcoming after climate shortcoming. For many in positions of power, the norm is insufficient attention, resolve, and real action. Too few setting strategy in corporate boardrooms, or setting down laws in chambers of government, seem to grasp the stakes of this moment, much less step up to meet them. Plenty are fighting to burn every last barrel of crude, come actual hell or high water.

In the written-spoken words of poet Andrea Gibson: "Why are the keys to our future in the hands // of those who have the longest commutes / from their heads to their hearts?"

So what is the antidote? The one that tugs insistently at my imagination is a *leaderful* upwelling. Like nutrient-rich water rising from the deep sea to the surface, I imagine a dynamic flow of people who care and commit to a safe, just, and lush world. If many more of us join in—with verve and cooperation—we could rectify the leadership crisis at the heart of the climate crisis and, just maybe, grow a life-giving future for everyone.* Even if that is a long shot, shouldn't we at least try?

In this fraught thicket of time, this is the pinch we feel: *There is no hope, and there must be hope. There is no way, and there must be a way.* We are caught between peril on one side and possibility on the other. If we are to navigate this nexus of distress and dreaming, we need contemplative, collective spaces to reflect, make sense, and then move together. Yet, more than ever, we seem to lack them.

A temple, synagogue, mosque, or church might be an obvious place to start, but fewer and fewer of us are part of one. Just 37 percent of American adults belong to a house of worship. (Admittedly, I am not one of them.) Even if we do, these institutions often remain unresponsive—or insufficiently responsive—to the realities of a heating, heaving planet.

*I first articulated the idea that "the climate crisis is a leadership crisis" in 2019 and proposed the need for approaches to leadership that are more characteristically feminine and more faithfully feminist. I further honed and unpacked those ideas in the opening essay, "Begin," of *All We Can Save.*

Classrooms are another place to turn, but we may find most are ill-equipped for our more-than-intellectual needs. In civic and movement spaces, emphasis on urgency and action can crowd out efforts to do this quieter soulful and relational work. (There are, of course, beautiful exceptions to these trends.)

We need to build or retrofit more spaces for holding our big questions, reverently and well, together.

> *What does it mean to be human, here and now? To help tend to a changing planet? To devote ourselves to the genuinely common good?*
>
> *What futures can we dare to dream?*

I have often felt responsible for providing a good answer to the echoing question so many of us hold: *What can I do?* I know well the agony that not-knowing can brew. But easy answers are phantom when it comes to a complex challenge. There is no simple formula, no fact sheet or checklist, for figuring out our roles in the vital work to forge a future we want and need.*

Navigating the twists and turns of my own climate journey, I have learned that rather than hunting for actions that are one-and-done or one-size-fits-all, a series of reflections and practices can help us work with that wondering—*What can I do?*—in beneficial and substantive ways. They can also help us work with questions down in the mycelium of the psyche. *What can I do?* is often the fruiting body of deeper inquiries about longing, belonging, and how to cope.

As we do that work, we can edge toward answers that are not about ticking a box but rather embodying wholehearted care for life on this planet. At times, that isn't about action, in its common trappings, at all.

*In 2018, as I was preparing to give a TED Talk, one of the curators asked me to leave the audience with a call to action. It prompted me to do my first writing (and speaking) about the true complexity of that request and the need to offer our superpowers, whatever they may be. A few years later, I published a five-step approach to answering the question of *What can I do?* That piece for *Time* was the foundational thinking for what became *Climate Wayfinding*.

Climate Wayfinding

Per dictionary definition, the term *wayfinding* means "the act of finding one's way to a particular place." In other words, sussing out where we are, working out where we want to go, and then finding passage to get there. It is a purposeful navigational act, one that has been fundamental, even defining, for the human species.

In Indigenous wayfinding traditions—like those of the Sámi, Polynesian, Inuit, and Aboriginal peoples—an array of signs and signals guide navigation through particular waters or across a given terrain. The location of stars above and the sun on the horizon. The movements of wind and ocean swells. The location of specific trees, rocks, and lichen on trees or rocks. The tracks of animals and patterns of snowdrift. Alongside environmental cues, oral history and song play a key role, as do memory, instincts, intuition, and even dreams. Colonization has eroded many of these practices, yet they persist, even as they now also bear the weight of climatic changes that scramble landscapes, seascapes, and skies.

Beyond humans, all species that migrate—from sockeye salmon and leatherback sea turtles to Arctic terns and humpback whales—are engaged in their own wild wayfinding. Seasonal rhythms, the Earth's magnetic field, and even smell guide the journey, which can span thousands of miles. They are endowed, it seems, with some sort of biological compass.

Attunement to place, and our position and movement within it, has also shaped entire realms of professional practice. In the field of urban design, for example, wayfinding entails the use of architecture and signage to help people orient and steer in a given locale. We build directional clues into human-made spaces.

Climate Wayfinding refers to the navigational process we are called into, as our home planet trembles with loss and change but also such potential. It is a lifelong practice, aided by intentional exploration, connection with our inner guidance, the support of others, and trusting in something greater than ourselves.

As with reading signs or stars *out there*, we engage in dialogue with the living Earth. We also go inward while keeping the big picture in view—a critical skill for this period of planetary transformation. We might think of *Climate Wayfinding* as a way of being in the world, rooted in sensing for *what is* and *what could be*, for relatedness with ourselves and other beings, for finding some steadiness in a shaky time, for making paths where there may be none.

> How does it feel to find yourself without a path? How do you greet the unknown?

To see more clearly the path ahead, we need to bring together three powerful and intersecting lenses. We need to *look inward with care, look outward with curiosity*, and then *look forward with courage.* In the simplest terms, these are the core practices of *Climate Wayfinding*.

We look inward to welcome and witness the essential *emotions*, core *motivations*, and guiding *values* that we bring with us into our climate engagement. We reflect on what insight or counsel they offer. And we discern and honor the *skills* or *superpowers* we want to contribute to collective transformation.

We look outward to scan the wide range of climate *solutions* and *accelerators* for change. We use our energy and curiosity to filter and gain focus. And we survey our unique *contexts* and *communities*, to consider where we want to engage and who our kindred collaborators could be.

We look forward with *vision* and weave together our threads of inquiry and discovery to create a personalized Climate Compass—a living navigational tool for the journeys we each take from here. And we plan next steps to deepen or enliven our climate engagement, taking *action* in our personal, professional, and public lives.

3 Core Practices: Look Inward, Outward, and Forward

In this moment and beyond, ours is a voyage without a clear destination. We know certain things we want for the climate-safe world we are trying to build. We want fossil fuels to be replaced by clean energy; buildings and cars to be efficient and entirely electric; cities to be redesigned for people over engines; waste to be rewoven into fresh material; ecosystems to be protected and restored; and regenerative food systems to flourish. We want all communities to thrive in and through this evolution.

But there is so much that remains opaque, still to be imagined and puzzled out. How will we see to the survival of people everywhere, and the more-than-human world, as we curb pernicious emissions, industries, and consumption?* How will we care for the hearts breaking and lives rent as losses continue to spread and sharpen? How will we evolve our shared stories

*Cultural ecologist and eco-philosopher David Abram coined the term *more-than-human world* in the mid-1990s. It is an invitation to turn our attention and appreciation to the myriad life-forms who also share this planet—to shed a sense that humans are so very special and that others, somehow, are not.

and beliefs and do so swiftly enough for the urgent metamorphosis we need? How will fairness and repair come to the fore?

We are immersed in a swirl of *yet*—but we can commit to finding our way with one another.

> What already gives you sustenance amid the swirl of *yet*?

Now in my forties, I still have the handbound journal that collected my teenage pangs and aspirations. Holding its worn pages in my hands, I wish I could somehow write back in time to my sixteen-year-old self and say to her, *Welcome.*

Welcome to being human on Earth. The ache you feel, the convictions you spelled out, the questions you are holding—they are all evidence that you are alive and a part of this wounded yet dazzling planet. Trust, if you can, that you will learn to let the questions work on and with you and begin to live life as a response. One answer, and then another.

The path will be winding, more the way of water than a straight-shot pine. There will be moments of feeling utterly lost, but also stumbling into awe. Along the way, you will come to know your own being and its marvelous interrelatedness at ever greater depth. Let yourself feel the firmness of Earth underfoot, dear one, as you step forward from here—mapless but journal in hand.

So very much has already shifted in my lifetime. Much more will in the rest of it. There is no easy bypass or relief road to steer clear of change.

We live in a liminal world. The Latin *limen* means "threshold"—literally the bottom of a doorway, crossed over as we enter a house or room. If we imagine ourselves at the point of entry, of beginning, we might give ourselves some grace for not knowing, for our angst and clumsiness. The verge can be an unnerving place. We might also recognize, here at the threshold, that we are being asked to bridge between the world as it is and the world as it could be. It can be hard to find our place and contribution in the space between. And yet, we must.

As Penobscot author and teacher Sherri Mitchell (Weh'na Ha'mu Kwasset) explains, there is incredible power, even magic, in that space. She uses the metaphor of the air held between a drum's round shell and taut skin, where vibration kindles sound. *Climate Wayfinding* meets us there—in the liminal, the emergent, the between. It accompanies us in the intermingled certainty and uncertainty of calling.

We will continue to face crossroads in the months and years ahead, and they will become more challenging as climate impacts swell.

> *Could our dismay and confusion call us into relationship?*
> *Could uncertainty beckon imagination and creative response?*
> *Could perception, pluck, and even play join us on the journey?*

My hope is that we will equip ourselves to greet the inevitable crossroads with more openness, more deftness, and more reciprocity; that we will grow our capacity to sense for openings, for next steps, though we may stumble as we go.

Climate Wayfinding is about finding our way forward and conjuring the future, yes. But it is also about remembering. We are each part of this sacred, extravagant community of life—alongside the dark tangles of rhododendrons, the unfurling fronds of ferns, and the rapture of moss in the rain.

We, and our gifts, belong here.

Equinox

Tamiko Beyer

Dear child of the near future,
here is what I know—hawks

soar on the updraft and sparrows always
return to the seed source until they spot

the circling hawk. Then they disappear
for days and return, a full flock,

ready. I think we all have the power
to do what we must to survive.

One day, I hope to set a table, invite you
to draw up a chair. Greens steaming garlic.

Slices of bread, still warm. Honey flecked with wax,
and a pitcher of clear water. Sustenance for acts

of survival, for incantations
stirring across our tongues. Can we climb

out of this greedy mouth,
disappear, and then return in force?

My stars are tucked in my pocket,
ready for battle. If we flood

the streets with salt water, we can
flood the sky with wings.

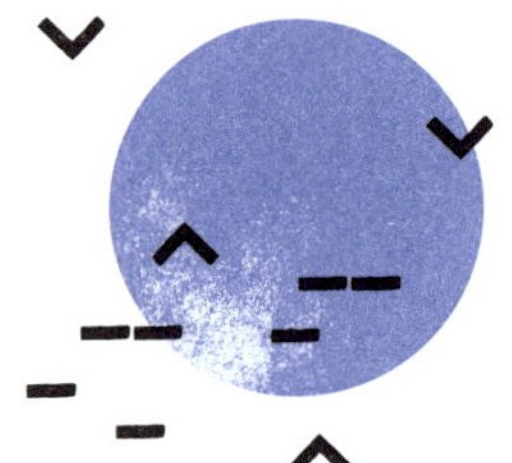

✡ Lighting the Way

with Colette Pichon Battle (Bayou Liberty, Louisiana)

In this time of planetary trouble and transformation, we need lanterns—wise souls who can help us make sense of where we are and how we might orient our lives. Here, we turn to Colette Pichon Battle, a leader in the global movement for climate justice.* She illuminates the power of heeding the call, however daunting, and charting an undaunted response.

Colette's roots are in the Gulf South, a place her people have called home for generations. A place permanently imprinted within her. A place that rising seas and storms are taking away.

I come from a Creole community in South Louisiana. Every Sunday, church is under these big oak trees that have been there for hundreds of years. The part of Spirit that's really free is out in those bayous, under those trees. These words that you learn—like "magnificent" and "holy"—you see them in front of you. I grew up walking miles in the woods. And that was safe because you didn't just belong to your family. You were the community's kids. All of that is sacred to me.

Colette did not grow up being afraid of storms. Hurricanes were part of life. They required a predictable dance, with steps her family and community knew well. But the heating of our planet has changed that.

When I was young, it was a hurricane party—literally, a party. All of your cousins, all in a house, and all your aunts cooking. The power is going to go out, so you have

*The term *climate justice* recognizes that the physical phenomenon of climate change is riddled with inequity. Those who have contributed least to the crisis are often most affected by it and have the fewest resources to cope. The wealthy countries, industries, and people who have polluted most—and used far more than their share of the Earth's atmosphere—have an outsized responsibility to curb the problem and support repair. The burdens created by climate change should be fairly shouldered, and the benefits of climate solutions fairly shared.

to get the food cooked. It's a buffet. In any moments of fear, you're with twenty people who love you. You're with generations of people who know exactly what's going on. We're bayou people. We don't run from storms. That's just not what we do. But those hurricanes are not the ones we're facing now.

Colette was called back home, and into the work of climate justice, by the cataclysmic 2005 storm we call Katrina—and the many social storms spurred by its aftermath. She was living and working as a lawyer in Washington, DC, when Katrina made landfall at the border of Mississippi and Louisiana in late August.

I was watching on television, and this storm was just way too big. I remember checking in and my mom saying, "I'm here, I'm fine." And I'm like, "Get out of there." My mom packed three days of clothes. She went to my uncle's house. And we lost everything. We had a very large tidal surge. My entire community went under about twelve feet of water. And we were on high ground. Everybody would park their cars there. But we went under. Everything went under.

By October, Colette was back in Louisiana. There, she confronted an utterly unfamiliar landscape in a place she knew so thoroughly, so intimately.

The destruction was unbelievable. Everything was gone. Houses were all in the water. And it was just death. I'd never smelled death like that. The salt water intrusion killed all the vegetation in the swamp. I remember driving and crying: Why am I not seeing anything? My parish lost two-thirds of its tree cover in that storm. As a kid, you would look at the sky and all you would see is trees, you know? And they were gone.

Two years later, Colette found herself in a room with researchers, showing time-lapse maps of projected Louisiana land loss. Still grappling with losing so much in the storm, she understood for the first time how much more loss would come—including her community in its entirety.

I couldn't believe what I saw. The experts are saying, "This is going to happen, no matter what." So suddenly you realize that we will lose this land, these communities. This place that I hold so dear and all the stories I'd been told from a very long time ago—all of those stories are going to go. Everything that I knew to be who I was—all of that is going to be lost. It's surreal. And that land, for people like me, was tied to our freedom. That land and the right to be there was the difference between being enslaved and not. To lose that—it felt in that moment that we would lose everything. I still cry about it. And I take drives now, just to make sure I witness.

In this aching, devastating discovery of loss to come, Colette began to walk a new and unexpected path. She started working to shift the laws that marginalize people and make them even more vulnerable in the face of storms like Katrina. She mobilized efforts toward fair, equitable disaster recovery and *energy democracy*—putting sources of electricity generation into the hands of the community, rather than industry.*

In place of systems of extraction, Colette and the "krewe" of her organization, Taproot Earth, are planting seeds of a world marked by abundance and liberation. All the while knowing that for too many places, we are already out of time.

To really, really admit that you understand what is happening to the planet, it will break your heart. If you don't cry deep, hard tears for the state of this planet and all of the people on it, you don't yet understand the problem. Once you get to that place, the only thing that can bring you out of that kind of darkness is belief in something greater than yourself. For me, it is that spiritual connection and understanding a greater purpose.

People ask me all the time, "If you know your land is going to go, why are you still fighting?" Well, we're going to lose everything, and my last name is Battle. What am I supposed to do? I'm supposed to fight. But I'm supposed to fight with tools that build people up, not tools that take people down and take them out. And that's

* *Energy democracy* is the aspiration to do more than transition to clean, renewable power. Today's fossil fuel energy systems are largely owned by corporations. The public depends on them, but has little to no role in decision-making. This alternative model seeks local control and ownership of a shared resource vital to our lives.

patience. That's care. All of those things that they taught you in Sunday school. This is a moral dilemma we're in. It's not a scientific one.

I was placed here, in this calling. Not as a lawyer, but just as a human citizen. And this is what I bring to organizing. Every now and again, the law shows up, but mostly it's the community-building and the joy and the love.

Colette's story reminds us that our agency in *Climate Wayfinding* is only ever partial. We may not even see ourselves on a climate journey until the realities around us suddenly insist. The aftermath of a storm. The possibility of parenthood. The transformation of an industry. The world is working on us, shaping the ways we might chart within it.

I am in awe of Colette's willingness to be called—to know that so much will be lost, to know loss so intimately, and yet stay in the work. That is courage that lights the way.*

*Colette's story is rewoven from a conversation she had with Krista Tippett for *On Being*. Hear more from Colette at climatewayfinding.earth. ⊘

Playlist

Songs for our first steps.

"Crossroads"—Tracy Chapman

"Still Sun"—Obongjayar

"Come Along (Edit)"—Cosmo Sheldrake

Tune in at climatewayfinding.earth. There, you'll also find a playlist of instrumental songs to accompany journaling and the like.

Journal

Pull out your journal and freewrite in response to these questions. You might set a timer for 5-10 minutes or play a couple of instrumental songs.

- Reflect on the roots of your climate awareness and concern. Can you identify a spark, a waking, a calling in? What happened?
- In this moment, what core question(s) do you find yourself holding about your own climate engagement? As you write, honor them as companions on the journey.

Consider sharing some of your writing with someone you trust. Notice how they respond. Notice how you respond to sharing with them.

Step Out

This is a step to take as an individual reader, beyond the page. You'll start with a small one, focused on supporting yourself on this journey. (*5–10 minutes*)

+ Identify or create a space that feels set apart for engaging with *Climate Wayfinding*—somewhere that supports you in being aware and attuned, porous to possibility. It could be a place inside your home or outdoors.
+ Consider bringing in some beauty, comfort, and connection to the rest of nature. Something as simple as a cloth, plant, or candle might do the trick.

Whenever you are ready to engage with the next chapter of the book, go there. Let that be a sanctuary for you during this experience. (If you are doing *Climate Wayfinding* as part of a group, you might also cultivate this kind of space for your shared time together.)

Gather

This is a 75-90 minute experience.

Tunes *(~10 minutes)*
Play songs from the playlist as your group gathers. (Or curate your own!)

Poem *(~5 minutes)*
Read "Equinox" aloud. Then, ask someone else to read it a second time. (It's wonderful to hear poems in two different voices.)

Opening Circle *(10–15 minutes)*
Welcome everyone. Express enthusiasm for the start of this collective journey.

Then, begin an opening circle with this prompt. The group lead should go first and model this; then "popcorn" to someone else. (If you don't yet know names, naming something someone is wearing can work well: "Cozy green sweater, over to you!")

> Share your name, pronouns, and a place that was meaningful to you as a child.

Group Agreements *(~5 minutes)*
Bring everyone's attention to the Group Agreements at the end of the chapter. Invite people to read them aloud, one by one. Then, welcome any questions, comments, or suggested additions to these edges for your collective garden. Finally, request affirmation from the group (e.g., a verbal "yes" or thumbs-up).

Reflection *(~5 minutes)*
Ask the group to take a few minutes to read through their freewriting in response to the journal prompts from this chapter—and jot down any additional thoughts that arise. It may be helpful to remind them of those prompts.

> Reflect on the roots of your climate awareness and concern. Can you identify a spark, a waking, a calling in? What happened?

> In this moment, what core question(s) do you find yourself holding about your own climate engagement? As you write, honor them as companions on the journey.

It can be nice to play instrumental music during this. After one song or a few minutes, invite people to wrap up their writing.

Discussion in Trios (*~15 minutes*)

First, introduce trios and how they work. They are a core part of the *Climate Wayfinding* experience, and a powerful one.

> We'll now break into trios to discuss what emerged for you.
>
> Each person will have 3 minutes to share from their own experience; then you'll switch. After everyone has the chance to share individually, there will be a final 3-minute round for any additional reflections before we come back together as a large group. If you're in a group of four, the last person will share during the final round.
>
> This piece is critical: When you're not the one sharing, your role is simply to listen generously, without offering response, commentary, or advice. You are just offering your attention. It may feel awkward at first, but go with it. We do trios this way to tap into the power of sharing from our inner wisdom and being witnessed in that sharing—with compassion and without interruption.
>
> For some of us, sharing like this might feel easy and natural; for others it might feel uncomfortable or difficult. You are the best judge of how and how much to share.

Check to see if anyone has clarifying questions. Then, invite everyone to form trios (you included!) to share from their journaling. Ideally, you have a way to keep time and can let people know when it is time to switch (e.g., ring a bell). Otherwise, ask the trios to manage that themselves. When you're ready, begin four 3-minute rounds.

Discussion as a Group *(10–20 minutes)*

Bring the whole group back together. Invite people to share what arose in their trios. You may not have time to hear from everyone, but it is nice to hear from at least one person in each trio.

You can also use these questions to spark or deepen the discussion.

> How do you experience "holding the questions"—being with the not-knowing, the uncertainty?
>
> What are you still mulling from this chapter that you'd like to push around together?

Closing Circle *(5–10 minutes)*

Express gratitude. Remind the group of when and where you will gather next time. (It's always best to handle any logistics before the closing circle.)

End with a closing circle using this prompt. The group lead should go first and model this, then popcorn to someone else.

> With a spirit of brevity, share one question you want to honor as a companion on this journey.

Closing Quote *(1–2 minutes)*

As a coda, read a passage from this chapter that inspired you.

Group Agreements

Generous

We will ask open and generous questions, offer our own stories and ideas generously, and listen to one another with a generosity of spirit.

Equitable

We will have a single conversation—one voice at a time, with roughly equal time to share. We will each step up or step back with awareness of our own proclivities and privilege.

Confidential

We will ensure that sharings made within our circle are not shared beyond it, unless someone gives clear permission to do so.

Growing

We will lean into learning, welcome diverse opinions and perspectives, and support our mutual growth in knowledge and power.

Courageous

We will bring our heads and hearts to this space, holding hard truths while looking toward what is possible and how we can best contribute.

Joyful

We will divest from perfection, performance, and posturing. We will welcome laughter, soak up the light when it comes, and celebrate progress and one another.

2

Getting Our Bearings

Pause in the present.
Then, pivot to the possible.

Earth's atmosphere is, among other things, a paradox. The invisible quilt of gases around our planet swaddles life, historically holding just the right amount of heat to nurture *Homo sapiens* and so much else on this teeming, humming sphere. O, praise to you, greenhouse effect.

But the balance is off due to pollution from industry—the equivalent of adding to the skies ply upon ply, fluff upon fluff. Now, that quilt, our vital buffering firmament, has grown thick with all those additional molecules of heat-catching, heat-keeping gas. In other words, the atmosphere is how we find ourselves alive at all, and also in this hot mess called climate change.

Given the scale and urgency we face, it makes some sense that we look to "big" levers for change—things able to make a dent in the roughly 50 gigatons of greenhouse pollution emitted globally every year. Exponential expansion of clean energy. Sweeping regulatory packages. Market-shifting movement of capital. To be sure, we need them, and fast.

However, in our yearning for technical fixes, miracle policies, and mass interventions, we sometimes overlook the most critical infrastructure of transformation: *people*. We are the living, breathing fiber of planetary healing, making all the rest possible.

Maybe it's no surprise that people, as a force for positive change, are consigned to an afterthought, time after time. As science identified the phenomenon of climate change and came to understand its contours, it also diagnosed the problem as *anthropogenic*. That is, "made by humans." The very term for the topic at hand puts an emphasis on the ills we create. (I can't help but hear echoes of Judith Viorst's classic tale of Alexander and his wretched day: *Terrible, horrible, no good, very bad humans.*)

Dominant cultural biases around how to effect change only make matters worse. Hard science is matter-of-fact. Humans? *Yeesh, they're messy*. Far better to stick with engineering and economics over ethics and emotions—to bank on mechanical, linear approaches. So focus goes to dollars invested, technologies deployed, and tons of carbon dioxide reduced. Those are "necessary." Human-focused efforts are relegated to the realm of boosterism and bake sales—harmless really, but merely "nice to have."

We stick to this peopleless path at great risk. Because if the problem of climate change originates in human activity, then planetary healing must as well. We must draw a different association between our species and Earth's conditions. The cultivation and continuation of widespread climate engagement is *itself* a vital solution—or, if you like, a crucial foundation to every solution, from telling captivating climate stories to installing rooftop solar systems. We will only move away from the brink if we center people in our grand U-turn.

> Have you ever felt shame around being human? Have you felt awe for people as a positive force?

So, we might ask: What would it mean to take seriously the conviction that each one of us can be a *node of possibility* for climate healing? And to tend that possibility in ourselves and one another? In other words, how do we realize the leaderful climate community we so need?

We have little research about what sparks and sustains climate engagement over time. Studies look at why people participate in a particular march. Biographies trace a single leader's path. I did my own doctoral research on what was then a burgeoning community of climate leaders within American evangelical Christianity.* But much remains a relative mystery. We know far more about the global climate system than the global climate movement and all those who comprise it.

At the helm of The All We Can Save Project—a nonprofit dedicated to nurturing climate leadership—I knew we would need some kind of framework to guide our work. If living buildings begin with renderings and blueprints, and ecological restoration jumps off from sketches and schematics, the *human* infrastructure of transformation deserves something similar. We need a model that centers people as essential rather than bypassing them. We need a model that speaks to the nodes.

*In 2012, I published this body of work as a book, *Between God & Green*. I wish, for the sake of American climate politics and our planet, that the subtitle had sustained: *How Evangelicals Are Cultivating a Middle Ground on Climate Change*. It became clear that religion could only carry people so far and was often cut off at the pass by political ideology—though some determined creation care efforts still continue within the evangelical community.

So, I dusted off my social science hat, drew on what research we do have, and supplemented it with many years of observation and my own lived experience. I synthesized a set of core dynamics that support and enable us to take part in climate healing over time. We call our novel framework the 8 Dynamics of Climate Engagement.*

8 Dynamics of Climate Engagement

* You can engage with an interactive version of the 8 Dynamics at climatewayfinding .earth.

As you see in the heart of the framework, the aim is clear: *deep, sustained, courageous climate engagement*. The links connecting the eight circles suggest that the strength of one dynamic buttresses others, and all are needed in an ongoing flow.

The statements are each framed from the perspective of an individual—*I have, I feel, I find*—to support self-reflection. It makes the framework personal. That "I" framing also honors that *systems change*—addressing the webby roots of a problem, not just tinkering with its symptoms—stems from what is small and close at hand.* But the real power comes when these dynamics hold true for many people: "I" upon "I" upon "I" as a multitude, all participating in the transformation in motion.

Each of the eight statements is succinct but carries layers of meaning.

1. I hold deep motivations to foster a just, life-giving future.

This is about purpose—the *why* or *whys* that fuel us, stoking the proverbial fire in the hearth of the belly. It is also about imagination—growing our capacity to look beyond the fissures and troubles of the present, which can sometimes be so all-consuming, and to catch sight of the future we *do* want, one that is genuinely life-giving for everyone, including the more-than-human world. Our motivations may reflect our values or guiding principles, and they offer us a wellspring for participation.

2. I feel connected to Earth and kindred community.

This honors the truth of our fundamental interconnection and interdependence. We *are* connected to Earth, to each other, though we may not always feel that way. Actively tending the felt nature of that connection—the vibrant threads between us—is vitally important. On so many levels, we need one another to

*The Systems Change Lab defines *systems change* as "the reconfiguration of a system, including its component parts and the interactions between these parts, such that it leads to the formation of a new system that behaves in a qualitatively different way." Think of building a new subway line instead of another lane on the gridlocked highway. Think of creating employee-owned companies, not just mandates to pay minimum wage.

take part in planetary healing, from cheering each other up to cheering each other on. And what is climate transformation ultimately about, if not returning to reciprocal relationship with the living Earth? We aim to move into alignment with life.

3. I can see myself in the unfolding, collective climate story.

Human beings are a storytelling species. Though it has not been central to much popular art, film, and literature—*yet*—the collective climate story is a tale being told in real time, in our time. Do we see ourselves in it? Do we trust that we belong in the narrative—not as bystanders, victims, or villains, but as actors in the most fundamental sense? Do we dare to take up the pen or paint a fresh scene or sing new verses? Ultimately, we are the shapers of this story, right from within its pages.

4. I'm able to work with climate emotions and access healing and rest.

The reality of the climate crisis is heavy, and climate engagement can be too. Around the world, there is swelling grief, anxiety, fear, anger, numbness, shame, exhaustion. Burnout within the climate community is all too rampant. In our rhythms as individuals and in groups, we need and deserve time to go dark and dormant, to pause and rest. Having the space, skills, and support to be with our difficult emotions is essential—to let the heart metabolize what is hard and to renew the wonder, reverence, and love that often live at the root of our participation.

5. I understand climate truth, just solutions, and leverage points for change.

This is about our ability to wrap our heads around the problem and grasp the ways we might respond to the urgency stamped in our atmosphere. When it comes to climate change, what are we dealing with, and how did it come to be? What solutions are in hand? How can we enact them in ways that are fast, fruitful, and fair, especially to those most climate-burdened? This is also about continuing to learn over time. We can keep following the questions that pull at our curiosity, honoring that knowledge is powerful and ever partial.

6. I have clarity around my contribution while embracing its evolution.

There is so much we each can do, and none of us can do it all. Not even close. Some of the work is more yours than mine, some more mine than theirs. As a small part of a vast collective, can we each navigate to our best ways to help? Can we hold clarity around what those roles are, while also allowing that they will necessarily evolve as we do—and as the world around us does? This is about moving through cloudy complexity into more lucid direction, again and again.

7. I find footholds for meaningful action and collaboration.

Here, theory becomes practice and intentions become deeds. To begin an ascent, a climber on a rockface reaches for a first handhold, a first foothold. Each hold allows for the next, though not always, or even often, in a straight line. Can we find ways to take meaningful climate action in our daily lives, at work or school, and in the public square? Can we find opportunities to collaborate, drawing on the might of many hands? Not every move will mean progress. It is a core capacity to stop, suss things out, and then continue the climb.

8. I have a sense of possibility, authentic power, and joy in the work.

This is really about *life force*—the most potent, persistent dynamic on this planet. Life force is our multi-billion-year inheritance, passed down to us despite all odds. It could also be our legacy. There is profound aliveness in possibility. Can we dance with the odds, even when they seem hopelessly long? Can we tap into our authentic power—not external power *over*, but internal power, rising up from within? And can we find joy along the way—delight in the very thing we aim to sustain, life itself?

> As you read through the 8 Dynamics, what are you drawn to? What sparks curiosity or excitement?

In the summer and fall of 2009, I was toiling to turn interview transcripts and focus-group findings into a dissertation. If you had asked me then to consider the truth of this framework for myself, I think I would have found the exercise fairly overwhelming. I was struggling with most of these dynamics. The lonely road of a doctorate had depleted me. Academia didn't feel like the right place, but where did? Climate policy was stuck in the mire. A global recession was making it hard to find a job. *Possibility, power, and joy? Are you kidding me?*

And yet, when I reflect back on that time and other personal crossroads—moments of feeling stuck, skeptical, troubled, or torn—I can see how helpful the 8 Dynamics would have been. I can imagine this framework validating my struggles and helping me identify what had become a hindrance or a hurdle to my climate engagement. It would have given me something for sensemaking, for getting my bearings, maybe even for figuring out a good step forward. More importantly, I would have known where I needed help—from my friends or mentors, my school or workplace—to transform those barriers and tend my small node of possibility.

Scanning the decades of my own experience as a climate feeler, thinker, and doer, I witness the ways these dynamics have flourished, waned, and revived. I can spot real low points, as in 2009, and I am certainly not always feeling strong on all of them. Some are easier or more intrinsic for me; others take real effort. But I can also spot moments of uplift, and, over time, the trends incline upward. The fluctuation will certainly continue—at times due to things that are out of my own control—but, today, the 8 Dynamics largely ring a bell of truth for me.

If a framework can have desires, this one hopes to be used. It's a philosophy that begs to be put into play. Every one of the 8 Dynamics benefits from space and support for reflection, connection, sharing, and renewal. Of course, a single pass through the framework isn't enough. These dynamics need ongoing care and conscious investment.

To that end, this is a tool to return to. It can shed light on our starting point when entering into something new, such as the journey of *Climate Wayfinding.* It can also make visible the shifts in our inner terrain if we come back to it at key junctures, such as starting a new year. We can use it alone or with others—within a class or team, for example. It can be a touchstone.

Ultimately, these dynamics nurture us in the wholeness of ourselves, as we confront the wholesale challenge of a planet under pressure. They open up pathways for our rich participation in climate healing, and from there, the overarching potentiality of a life-giving future.

> When you're feeling bewildered or adrift, what helps you to get your bearings and return to possibility?

Possible—that mainspring word in *Climate Wayfinding*—has a somewhat unusual etymology. (As you may have gathered by now, I am quite keen on the often-illuminating origins of words.) It comes from the Latin *posse*, meaning "be able." But the suffix, *ibilis*, then seems to double down. Meaning "able, or worthy, to be," it points to something one can act on.

Put the two together, and the word takes on a kind of cyclical rolling forward, almost train-like: *able . . . to be . . . able . . . to be . . . able . . . to be . . .* It's as if possibility wants to become reality, if we are willing to take part in the momentum. It's a summons to step into the locomotion.

But *possible* packs yet more punch. The Latin root also gives us our modern word *posse*, as in "a group of people." So I hear in it not just a summons but an axiom, too: *We* are where possibility lives.

Yes, we are. The 8 Dynamics framework uplifts that *we*—strengthening each filament, each confluence, each offshoot, and the whole of our quickening collective. With these dynamics at work, we can cultivate our innate human capacity to face the climate challenge and pledge ourselves to what is, despite it all, still possible.

So let us pause to offer praise, too, for all the people in the greenhouse. All of us looking up at the atmosphere with both awe and angst. All of us trying to put things to right.

I often draw back from the question incessantly posed to climate leaders: *What gives you hope?* What gives me a sense of *possibility* feels so much more tangible, truthful, real. But call it what you will, the source, for me, is clear: scores of people stepping into their own generative capacities and trying to make *anthropogenic* not terrible-horrible but a very good thing, in the end.

♥ The Big Picture

Ellen Bass

I try to look at the big picture.
The sun, ardent tongue
licking us like a mother besotted

with her new cub, will wear itself out.
Everything is transitory.
Think of the meteor

that annihilated the dinosaurs.
And before that, the volcanoes
of the Permian period—all those burnt ferns

and reptiles, sharks and bony fish—
that was extinction on a scale
that makes our losses look like a bad day at the slots.

And perhaps we're slated to ascend
to some kind of intelligence
that doesn't need bodies, or clean water, or even air.

But I can't shake my longing
for the last six hundred
Iberian lynx with their tufted ears,

Brazilian guitarfish, the 4
percent of them still cruising
the seafloor, eyes staring straight up.

And all the newborn marsupials—
red kangaroos, joeys the size of honeybees—
steelhead trout, river dolphins,

so many species of frogs
breathing through their
damp permeable membranes.

Today on the bus, a woman
in a sweater the exact shade of cardinals,
and her cardinal-colored bra strap, exposed

on her pale shoulder, makes me ache
for those bright flashes in the snow.
And polar bears, the cream and amber

of their fur, the long, hollow
hairs through which sun slips,
swallowed into their dark skin. When I get home,

my son has a headache and, though he's
almost grown, asks me to sing him a song.
We lie together on the lumpy couch

and I warble out the old show tunes, "Night and Day" . . .
"They Can't Take That Away from Me" . . . A cheap
silver chain shimmers across his throat

rising and falling with his pulse. There never was
anything else. Only these excruciatingly
insignificant creatures we love.

✡ Lighting the Way

with Dr. Kate Marvel (Brooklyn, New York)

This time calls for new cartographers—those who can map the physical transformations of our planet and decipher what they mean for human beings. Dr. Kate Marvel is uniquely adept at both. As a climate scientist, she models the potential pathways ahead for her "favorite planet," Earth, and all of us Earthlings. She helps us understand what it would mean to inhabit the realities rendered and, more importantly, grasp our chance to chart toward the better worlds, still reachable.

Most days, you can find Kate immersed in the equations and code of climate models or writing her take on the collective climate story. She was born in California and grew up smack-dab in the center of Ohio—a kid who was more fascinated than most by the nature of reality. But she never once considered being an Earth scientist. In fact, the creepy scientists in the film *E.T.* had given Kate the distinct impression that the whole field was terrible.

My path took an unexpected turn in undergrad, when I took Introduction to Astronomy to fulfill a course requirement. I remember sitting there and thinking, "This is amazing! Why didn't anyone tell me this was amazing?" I thought everything we were learning was so cool and wanted to know more, so I switched my major and then continued to study physics all the way through to a PhD.

But that was *theoretical* physics, which can tend toward the abstract, if fascinating. Think: string theory. Kate cares immensely about the material here and now. She didn't want to run thought experiments; she wanted to understand and inform the real-time experiments playing out on our planet. She wanted to answer questions that matter, deeply, for what we might then do.

I drifted through different topics for a while. Then, I found climate science. Lots of physics, computer programming, data analysis, and all of it dealing with real-world problems—maybe the *real-world problem.*

My colleagues and I build virtual Earths—mathematical representations of our planet's climate system—which help us understand what has happened in the past, what is happening now, and what could happen in the future. And we run experiments: What would occur if we double the amount of carbon dioxide in the atmosphere? Or, what if we stopped burning fossil fuels—coal, oil, gas—tomorrow?

I always come back to physics to get my bearings. You can write down an equation, and that gives you something to hang on to when things get complicated—and they always do.

The models Kate and other scientists build aren't perfect representations of the actual, intricate Earth, but they show us a few things with certainty. First, climate change *is* anthropogenic. We know without a doubt that natural phenomena are not causing the rapid heating on our planet—human-generated greenhouse gases are. Second, that change has happened with unprecedented speed in the sweep of known history. We have shifted rapidly from the climatic stability of the last ten thousand years—the Holocene's just-right "Goldilocks" conditions—into soaring instability, particularly over the last fifty-ish years and counting. Third, and thankfully, a range of possible futures is still available to us.

Digging up dead stuff and burning it. Chopping down forests. These are the main things that have catapulted us into this new disjointed age, often dubbed the Anthropocene. We're already seeing the difference that 1.4 degrees Celsius (2.5 degrees Fahrenheit) of global warming above preindustrial temperatures makes: more record-shattering heat, more severe hurricanes and thunderstorms, more droughts and floods. Almost across the board, these shifts are more significant than scientists anticipated. All of it injects more chaos into the world, creating new threats and multiplying existing ones.

I say this as a committed resident of this planet: We really don't want to live in a world of 1.5 degrees Celsius (2.7 degrees Fahrenheit), much less 2 or 3 degrees Celsius (3.6 or 5.4 degrees Fahrenheit)—and somewhere between 2 and 3 degrees Celsius is the path we're on right now. Every tenth of a degree of warming matters.

But the same science that shows greenhouse gases are responsible for climate change also shows that when we stop emitting them, the global temperature will stabilize. How bad it gets is up to us. The future, and our safety, is still in human hands.

As Kate works to answer big questions about and for Earth, she immerses in all that remains uncertain in climate science: whether the vast tropical forests of the Amazon will turn into dry savanna; whether the West Antarctic ice sheet will collapse and hoist sea levels not by inches but feet; whether the crucial Atlantic Ocean overturning circulation will shut down and chill Europe. She must continue to grow her own capacity to be with the uncertainty that is inherent to being alive at this time.

I'm a scientist, which means I don't believe in false hope. But I also don't believe in false hopelessness. We can absolutely reduce the suffering that climate change will cause. The task is very clear: reduce greenhouse gas emissions as much as possible, as quickly as we can. We have most of the tools we need to do it.

The biggest variable in determining our climate future isn't some physical process we don't understand. It's us. It's human behavior. It's what we collectively decide to do, connected in a web of concern and action. And that's actually empowering when you think about it. It means we have agency.

I want there to be options beyond just calamity, other stories to tell. It will not be easy, but there's a possible future where we can look back and pat each other on the back, feeling proud of a job well done.

Part of the cartography of climate science is this: Even if we follow the *most* optimal climate trajectory, we, Earthlings, are in for rocky terrain, and rising seas, ahead. It's essential for us to grasp that so we can respond. But Kate maps something else, too: Even under duress and amid so much loss, Earth is precious.

Everything I love—and everything I will ever love—is here on Earth. This is the best place in the universe.

At its core, cartography is storytelling. There's a conceit floating around, of late, suggesting we skip town and rocket away. Kate reminds us that's pure nonsense and calls us to something so much better: to get our bearings right here on this watery planet, fragile yet resilient, and make of the Anthropocene something beautiful. I imagine light feet. Linked arms. Voices lifting, maybe in song.*

*Kate's story is rewoven from an exchange we had, as well as other material. Hear more from her at climatewayfinding.earth.

Playlist

Songs for finding our footing.

"I'm On My Way"—Rhiannon Giddens and Francesco Turrisi

"Raíz"—Bomba Estéreo

"Matter of Time"—Sharon Jones & the Dap-Kings

Tune in at climatewayfinding.earth.

Map

Use the 8 Dynamics of Climate Engagement to create a map of where you are at the beginning of this journey. (*5–10 minutes*)

What you'll need: the interactive 8 Dynamics Quiz, found at climatewayfinding .earth. (If you aren't able to go online and use the quiz, capture your answers in your journal.)

+ Take a moment to reflect on each statement and consider what is real for you, right now.
+ For each dynamic, mark your response from 1 (*that's not true for me*) to 5 (*that's extremely true for me*).
+ After taking the quiz, save your final "web." (You'll want to look back at it later in the journey.)

Journal

Based on your 8 Dynamics map, do some freewriting on the first prompt.

> Within the 8 Dynamics, what feels wobbly? What feels strong? Pick two that feel most significant for your own deep, sustained, courageous climate engagement. Consider how you might invest in them with intention.

Now, read the prompt below, and take a minute to close your eyes, breathe, and see what appears in your mind. Then, put pen back to paper.

> If this era of human-made change were to become something beautiful, what could that look like? What would people do and be?

Step Out

Philosopher and activist Simone Weil called attention "the rarest and purest form of generosity." With this step out, you'll generously turn your attention to the more-than-human world. (*20–30 minutes*)

+ Find a place to be outdoors. Spend some minutes simply observing the lay of the land and the life there: stillness and movement, independence and interconnection. You can do this sitting still or meandering slowly. If you'd like, jot notes or sketches.
+ Then, forage for wonderings about this place. What are you curious to understand so you can relate to it more deeply? Perhaps aspects of ecology, geology, hydrology, history, culture, inhabitants present and past, challenges faced, and what may lie ahead.
+ After you've focused your attention outward, bring it inward. Notice how it feels to be you in this place. Scan your body for sensations. See what stirs within you.

Following your time "in the field," you might consider research, writing, making art, or all three to enrich or express whatever you found there. You might also return to this place throughout your *Climate Wayfinding* journey—notice the movement there, even as it stays put.

Gather

This is a 75-90 minute experience.

Tunes (*~10 minutes*)

Play songs from the playlist (or DJ's choice) as your group gathers.

Poem (*~10 minutes*)

Read "The Big Picture" aloud. Then, invite someone to do a second reading.

Opening Circle (*10–15 minutes*)

Welcome everyone back. Then, begin an opening circle with this prompt. The group lead should go first and model this, then popcorn to someone else. (Sharing names again will help them take root.)

> Share your name and something human-made or human-done that reminds you anthropogenic can be a good thing.

Reflection (*~5 minutes*)

Ask the group to take a few minutes to read through their freewriting in response to the journal prompts from this chapter—and jot down any additional thoughts that arise. You might remind them of those prompts.

> Within the 8 Dynamics, what feels wobbly? What feels strong? Pick two that feel most significant for your own deep, sustained, courageous climate engagement. Consider how you might invest in them with intention.
>
> If this era of human-made change were to become something beautiful, what could that look like? What would people do and be?

After one instrumental song or a few minutes, invite people to wrap up their writing.

Discussion in Trios (*~15 minutes*)

Invite everyone to form trios to share from their journaling. Ideally, you have a way to keep time and can let people know when it is time to switch (four 3-minute rounds). Otherwise, ask the trios to manage that themselves.

Discussion as a Group (*10–20 minutes*)

Bring the whole group back together. Invite people to share what arose in their trios. You may not need to offer any other prompts, but you can also use these questions to spark or deepen the discussion.

> Within the 8 Dynamics, where do you feel well-resourced and equipped to support others?
>
> What insights resonated with you from this chapter? What gave you pause?

Closing Circle (*5–10 minutes*)

After a reminder about your next gathering, end with a closing circle using this prompt. Model it yourself; then popcorn.

> Share a word or short phrase that you'll carry forward from this second gathering of our group.

Closing Quote (*1–2 minutes*)

As a coda, read a passage from this chapter that inspired you. (Or! You could ask someone in the group to prepare a selection in advance and read it.)

3

Following Our Emotions

Look in on the feelings
that surge, eddy, and send you on.

Have you ever found yourself awake at night with the climate crisis, and its intersections, twisting through your mind? I certainly have been there.

Tired eyes wide-open to the immensity of our planetary challenge. Grieving for what has already been lost, and fearful of the losses to come. Outraged—we did not have to be here. Hopeful—humanity could still choose a better path. Despairing—I am not sure we will. Working and working the Rubik's Cube of the untenable status quo. Wondering if the ways I am trying to help shift it are, in fact, helping. The truth is, if you are awake to the climate crisis during the day, finding yourself awake at night is a distinct possibility.*

There is no getting around it: What is unfolding on our planet brings with it big feelings. Between October 2020 and October 2021, Google searches for "climate anxiety" soared 565 percent—a simple online act that offers a profound glimpse into a collective psyche. More recently, nearly half of young people surveyed said their feelings about climate change have a negative effect on their day-to-day life and functioning. How we feel about our planetary trouble can impact our ability to be engaged students, friends, colleagues, or members of a family or community.

With more and more people encountering climate impacts directly—storms, fires, floods, and more—all of this multiplies in prevalence and intensity. When we encounter the climate crisis up close, despair, panic, anger, numbness, or depression can be an outgrowth of firsthand trauma. Increasingly, entire communities face such shock and strain. What Dr. Lise Van Susteren, a practicing psychiatrist, termed *pre-traumatic stress* can seize those who know climate loss mostly from afar, through journalism or social media or anticipation of what may lie ahead.

As with all aspects of the climate crisis, disparity and inequity are present in its emotional and mental health dimensions too. For some, this is a first encounter with what feels like the end of the world; for others, it is yet another wave of worlds ending.

What is playing out on the surface of the planet is reverberating inside us in many different ways. And of course it is. Entwined as we are with Earth's living systems, that makes sense, doesn't it? We are knitted together with more than eight billion human beings and over two million known species.

*In 2021, Dr. Britt Wray and I collaborated on a curation of seven resources for coping with climate anxiety, published by *Time* magazine. I wrote some of this as framing for that piece.

We can feel the connections between us. We can also feel the threadbare places within the fabric of life, the slipped and fraying edges—and the closing window to mend them.

> What emotion stirs within you at this very moment? Can you greet it, whatever it is? Maybe say *hello*.

As climate and mental health expert Dr. Britt Wray reminds us, climate anxiety—or sadness or indignation or shame or helplessness—should not be dismissed as catastrophizing or overreacting. It is a healthy response to an existential threat, one that has implications for our lives and all that we love. In other words, our felt reactions are profoundly human. It might be that the wounds themselves are what bring us to *Climate Wayfinding*, as they have a way of sending us searching.

Dr. Panu Pihkala is an interdisciplinary researcher on eco-anxiety. Through a trailblazing "taxonomy of climate emotions," he has sought to put form on the amorphous, splayed array of emotional reactions to the climate crisis, of which anxiety is just one. That taxonomy gave rise to a map of sorts—a wheel of climate emotions—co-created with journalist Anya Kamenetz and Sarah Newman of the Climate Mental Health Network. (See the Climate Emotions Wheel on the next page.)

As this wheel shows, our climate-related emotions are not exclusively feelings of distress. Interest, inspiration, gratitude, and the like may also spring up. But most of them are uncomfortable, at the very least. No wonder, then, that it can be tempting to push these feelings away.

I have found that inclination to be especially true when supportive spaces to surface, share, and work with climate emotions are so few and far between—and when many of us have not even considered that there could be such spaces to heal collective trauma. Casting into the void of Google can seem, at

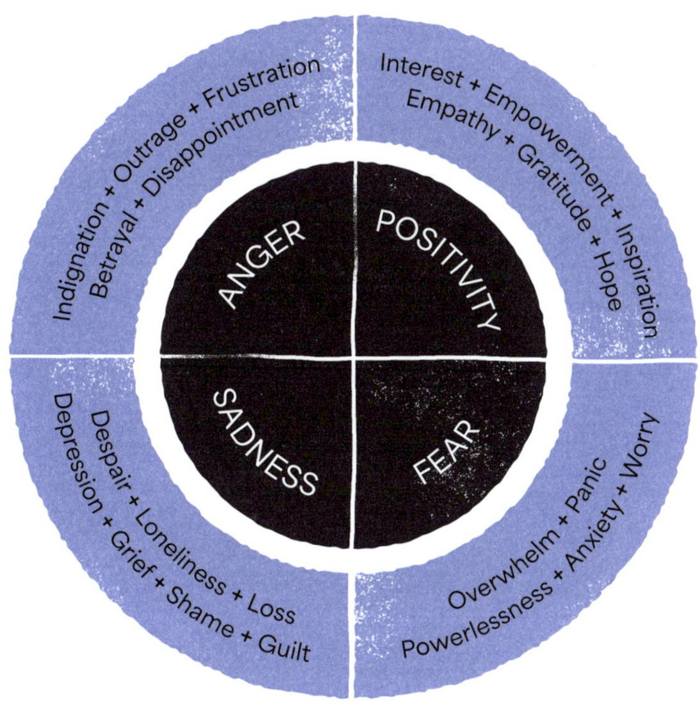

The Climate Emotions Wheel

times, like the best and only option we've got. We may find ourselves asking, as I did in a poem once: *Is there somewhere soft to kneel, to keen, beneath a freighted sky?*

Grief tends to be the climate emotion I feel most often and most deeply. Every day there is climate news that will break your heart: another code-red report, another record broken, another irrevocable extinction, another teetering on the edge, another falling, falling short. Hell, it can be as simple as a warm day in January—beautiful out of context—or strawberries for sale at the farmers market a month early. Messengers of an eroding Earth are all around us.

As Dr. Kate Marvel writes in *Human Nature*:

> *The planet, like us, has lost much before. But never quite like this. This loss is everywhere, it's fast, it's inevitable. It's sad in a way I can't describe, an inaudible crack in an invisible surface, the breaking of billions of hearts all at once.*

If there is nowhere for it to go, the grief lodges in me, settling heavy into my bones and impeding imagination or action. I feel like I am straining against some innermost gravity. Grief can also take other forms. It seems to have a way of operating in disguise at times, wearing cloaks from defensiveness to rage. But neither grief nor its familiars have a permanent claim. I have learned that I hold sway as to whether my heart simply breaks or breaks *open*.

> Do you have a primary climate emotion? How does it reveal itself in your flesh and bones?

Over the years I have gotten better at finding, or making, spaces of solace and healing. I call on guided meditation to be with difficult emotions, and I let my body move as it wants to express them. Sessions with a caring climate-aware therapist help me share, feel, and renew. A spiritual practice helps me connect to something bigger, more enduring; so does lying down outside with a dear friend, Earth at our backs, gazing up at an awning of trees and clouds rambling above them. Sometimes, I just put my hand to my chest and let the tears come, like clouds admitting the secrets they had tucked away.

Connecting with caring community, even a community of just two people, can make all the difference in my capacity to cope. For well over a decade now, I have gathered, monthly, with a heartful group that gives me a sense of harbor, where I can take whatever is rumbling in the hull of myself. I find truth and guidance in the words of poet-essayist Adrienne Rich: "There must be those among whom we can sit down and weep, and still be counted as warriors."

Climate Wayfinding intentionally creates space for looking inward and feeling our climate-related feelings—whatever they are and to whatever degree we wish to do so.

At its core, the journey of this book is about what being human means, here, now, on this hot planet. To that end, we welcome the questions we are holding and honor them as sources of direction and even sustenance. They are valuable companions. We also, and equally, welcome and honor the essential emotions we bring with us to this topic and experience, painful though they may be. We do that because these feelings don't sit at the periphery of our climate engagement; they are at its heart, quite literally.

For better or worse, our feelings drive how we show up in the world. Emotions manifest in our minds, as well as in our bodies and behavior. They seep into our words and mold our manner.

When we don't tend to our climate-related feelings, it can thwart the work we are trying to do. For example, we can be so inundated with distress that we find ourselves immobilized, unable to participate in the planetary healing that is still possible. *It's just too hard to care.* Or we might keep moving but find that our emotions discharge in ways that are detrimental—such as smoldering anger that pushes people away or pent-up frustration that undercuts every idea as too meager or too foolish. Our nervous systems, unattended, can get the best of us.

In 2022, I interviewed Dr. Kritee Kanko for an episode of the podcast *A Matter of Degrees.** She is both a climate scientist and a Zen Buddhist priest who facilitates ceremonies around our "exiled emotions." Echoing one of her teachers, activist and author Joanna Macy, Kritee told me bluntly: We have to "fucking face our grief," or else we will take "shallow actions." If we don't go into the depth of the pain, we will opt for Band-Aids or quick fixes or feel-good

*I co-host *A Matter of Degrees* with political scientist and climate policy expert Dr. Leah Stokes. Find this episode, "How to Cope with All the Climate Feels," at climatewayfinding.earth.

gestures that simply do not create the real change we need. We will tinker, rather than transform.

Our feelings, then, can keep us frozen or frittering, or they can become fuel. Writer-philosopher Audre Lorde puts it plainly: "Our feelings are our most genuine paths to knowledge." They might also be our most genuine paths to action, if we examine and follow them consciously—however chaotic or contradictory or throbbing they can be. "This is how new visions begin," Lorde says, "how we begin to posit a future nourished by the past."

Sherri Mitchell (Weh'na Ha'mu Kwasset), whom you met earlier, frames it like this: Our difficult emotions about our ecological and social plight are not a sign of something wrong with us, but a sign of something being *righted* within us. The first time I heard her say that, I experienced the release that comes from feeling seen by clear, kind eyes. And I feel it now in my body as vertical alignment—alert, upright, ready to heed Earth's call that a better way of being must emerge, and fast.

That is what I have learned, most of all, about a broken-open heart: It is awake and alive and calls for action. It is regenerative, like nature, reclaiming ruined ground, growing anew.

With a broken-open heart, I am not stuck in grief—or any other difficult emotion swirling inside me. I can access compassion, daring, and zeal. I can hold the simultaneous truths of destruction and possibility, tuning in to the movements, real leadership, and true solutions that are rising. I can return to the awe and reverence I feel for this magnificent planet and the ferocious love that lives at my core, a blend of tenderness and fire.

> Consider a place or being that kindles ferocious love in you.

Action, we are often told, is the antidote to despair. To be sure, research indicates it can have a beneficial effect on our climate-related mental health. Participating in collective action—*doing* in community with others—can buffer climate anxiety and build a sense of agency. But as someone who does climate work just about daily, I can personally vouch that that kind of action alone is not a panacea—though I find it much better than the opt-out alternative. (Tried that once, years ago; really ran aground.)

I realize the task lists are already long and the bars already so very high, but coping, caring, and healing is an essential part of climate engagement too. Our emotional capacities are just as important as the technical, strategic, or political skills we might employ. And the latter often cannot function well without them.

Here, the architecture of trees offers us something of a parable. As onlookers, we admire their trunks and barks, branches and leaves, casting their canopy spells. But that is only half the story. The other half takes place below ground, with root systems that can be a near mirror of the fanning out above. Extraordinary, productive, magical things happen *sub terra*. Trees are only able to reach for the light because they also live and thrive in the dark.

That's true for us as well. The parts of ourselves that are in full view, growing and swaying in the open air, depend on our emotional-spiritual roots, stretching down and out. They give us, like trees, strength to make the most of spring when it arrives—the next opening for climate engagement. Our roots can plait with others' in a system of mutual support. And when the next storm comes, which it will, we might discover we're more prepared to meet the torrents and the gales.

As the physical terrain of Earth continues to shift, so will the topography of our climate-related emotions. *Motion* is most of the word, after all.

The metaphor of a pendulum reminds me that emotional oscillation is simply part of what it means to be alive. The pendulum of the heart does not stay in one place. It swings into darker, heavier feelings. It swings into lighter, brighter feelings. It finds some breathing space in OK-ness and calm. I find some reassurance in knowing that I will not get stuck. By design, a pendulum is made to swing. And if it's made to swing, maybe that means it's just fine to feel whatever I am feeling—it's OK to be me, no matter what emotions are circuiting through. I can be with them, right here.

How many times in my life have I recoiled from something surging inside me? Whispered urgently, *No, stop, I don't want to feel this,* and tried to turn away? Countless, certainly. Chances are, just earlier today. But when I am able to greet my emotions with allowance, get curious and compassionate, breathe with them—then I find they often have much to teach me.

Feelings have a unique power to break through the din of daily life—all that chatters for my attention now, now, now—and to help me listen more intently for what is true. *Bow down in reverence,* they remind. *This here is precious,* they declare, *and needs protecting.* My feelings offer critical grounding, guidance, and sometimes prodding I may not want but probably need. *Get still and quiet. Make time to compost. Follow what you love. Come out of isolation and into community.*

For what if this grief or rage or despair is not mine alone? What if it's something shared by many, even across generations, even across species? What if it's another reminder of how inextricably linked we are, looped into the thin layer of life that swathes our spinning planet? And what if that sense of being so thoroughly and completely entwined is the deepest motivation of all?

Nothing Wants to Suffer

Danusha Laméris
after Linda Hogan

Nothing wants to suffer. Not the wind
as it scrapes itself against the cliff. Not the cliff

being eaten, slowly, by the sea. The earth does not want
to suffer the rough tread of those who do not notice it.

The trees do not want to suffer the axe, nor see
their sisters felled by root rot, mildew, rust.

The coyote in its den. The puma stalking its prey.
These, too, want ease and a tender animal in the mouth

to take their hunger. An offering, one hopes,
made quickly, and without much suffering.

The chair mourns an angry sitter. The lamp, a scalded moth.
A table, the weight of years of argument.

We know this, though we forget.

Not the shark nor the tiger, fanged as they are.
Nor the worm, content in its windowless world

of soil and stone. Not the stone, resting in its riverbed.
The riverbed, gazing up at the stars.

Least of all, the stars, ensconced in their canopy,
looking down at all of us—their offspring—

scattered so far beyond reach.

✡ Lighting the Way

with Sister True Dedication (Loubès-Bernac, France)

Today, we are not only walking the outer terrain of a changing planet. We are also walking the inner terrain of our emotional responses to it. Sister True Dedication (Chan Hien Nghiem) helps us see the potential that lies in the interlacing of those paths. Indeed, we cannot "fix" the outer without healing the inner, nor can we "heal" individually if we neglect society's broader needs. To privilege one while disregarding or sacrificing the other is not headway, not really. The true opening is the braided steps we take.

Before she became a monastic in the Plum Village tradition, Sister True Dedication was a young journalist working for BBC News in London. Her beat in the early 2000s was politics, and her hope was to use the media to effect change. But the newsroom, she found, was fraught with arm's-length indifference to people's suffering, even to their slaughter.

This was during the beginning of the Iraq War. We had banks of screens with live feeds of people dying in real time. But these men, my colleagues, would watch and laugh, like it was a computer game. Everything about it felt wrong to me. One day I took a deep breath, walked across the newsroom to where they were standing, and said, "What the hell do you think you're doing? You should be ashamed of yourselves."

I knew it would compromise me professionally, but it was so important for me to be a truth teller. It was an awakening: I wanted to live my values 100 percent. I didn't want them to be compromised. And I realized that toxic culture was creeping into me—that if I stayed, I would become it.

At that time, Sister True Dedication had already encountered the work of Zen Master Thich Nhat Hanh ("Thay"), a global spiritual leader, poet, peace activist, and founder of Plum Village, where she had already been on a first retreat. She credits the community's ethical code as part of what emboldened her to speak up that day in the newsroom. Shortly thereafter, she went on retreat again, then continued to return to study with Thay.

I thought, this guy is an order of magnitude wiser than every single one of my university professors. I feel like he has the answers to all of my questions. I can't not *learn from this person. During my three years at the BBC, I used all of my vacation days to go to Plum Village.*

Retreats eventually turned into residence there. Then, in 2008, Sister True Dedication committed herself fully to the path, ordaining as a Buddhist nun.

I wanted to live leanly on this planet. I wanted to die having served this planet well. And I knew that I needed to transform and heal a lot more, myself, before I could truly be of service. Here was a path of training to become a warrior of the soul, to be the light that we have within.

Daily life at the monastery is threaded with practices that cultivate mindfulness and brighten the spirit: mindful breathing, sitting and eating meditation, deep relaxation. Perhaps nothing is more central than walking meditation.

It's such a simple thing, this practice of walking meditation—to be aware as we walk and to be nourished by our steps. It's a way of just spending time with this planet that we want to help. The Earth wants to be enjoyed.

We have a beautiful dream that one hundred years from now, young people can walk through a forest and enjoy it. But if we, already, lose the art of doing so, then a hundred years from now, there's no chance. We must see the connection between our way of walking and what we are transmitting to future generations.

We don't need to be Buddhists ourselves to benefit from these practices and modalities. For those of us who want to change the world, we may find that shifting our consciousness is the first step.

It is so natural to want to eliminate injustice and suffering. But the way out, truly, is to train in how to respond to that suffering with fierce, all-embracing compassion. That is our mission as humans. What we are called to in this moment is levels of compassion we can barely believe, and, as Thay taught, if there's no contact with suffering, compassion doesn't have a chance to arise.

Sister True Dedication also foregrounds the opportunity to walk with our own difficult emotions through loving presence, with or without actual footfalls.

If we find in ourselves sadness, fear, anger, or despair, we have a chance to accompany it, to be with it, and to look at what it is telling us. Our suffering contains within it the suffering of our world—the violence, destruction, discrimination, and exploitation. The more we awaken to this, the more we want to do something to help.

Like roses that grow from a heap of kitchen scraps, in our suffering there is raw material for cultivating the courage to act. As the Zen teaching goes: *No mud, no lotus.*

Community also plays a role in transmuting pain. Sister True Dedication calls the sangha of Plum Village a "healing hospital." When we hold pain collectively, rather than shouldering it as individuals, we can create a different kind of emotional-spiritual alchemy.

In our tradition, we say that only when we can name the pain, be with it, and embrace it with mindfulness and compassion—only then can healing take place. Our collective energy can hold huge amounts of pain. We follow the breath. We let the tears roll. We surrender our bodies to the Earth; she can handle all of it. When we're willing to walk through pain together, then insight emerges.

I think we should sing from the rooftops that healing is possible. According to impermanence, everything is possible.

At Plum Village, Sister True Dedication returned to the written word, but in a very different way than she had as a young journalist. She began to edit Thich Nhat Hanh's books on Buddhism and ecology, including *Love Letter to the Earth* and *Zen and the Art of Saving the Planet.*

As a dharma teacher, she linked arms with other monastics and global climate leader Christiana Figueres to turn the latter book into a dynamic learning journey—one that could serve people struggling with the very ache that catalyzed her own spiritual path. In the closing session of that course, she draws our attention to the places where inner and outer terrain meet.

Our short lives are a field of action. How are we walking? How are we able to walk? Rather than striving and straining, we can truly walk on the ground of reality as someone fully present, free, and awake—someone responding to challenges with peace, compassion, and non-fear. Steps taken like this can heal the body, mind, and planet. Let us act as an expression of the stream of life.

Sister True Dedication helps us see that our way of being is a profound space for reverberating climate engagement. While we, and our contributions, might feel too small to matter, we can touch into infinite resonance in the very ways we walk through the world.

The planet is calling for radical, decisive action. We don't know if our actions will have the outcome that we want. But we can offer them, unconditionally, as prayer, and know that our way of acting is rippling out, even when certain outcomes seem to be failing.

Sister True Dedication is herself an embodied example of continually composting the overwhelm, bitterness, or anguish that can spring up so that healing action can be fueled by care, love, and delight in this Earth. That kind of motivation, she says, comes from a more clear place in the heart; it offers a wellspring of energy, replenishing to our depths, and a connection to something so much bigger.

Each one of us needs to identify and nurture our deepest aspiration, rekindle it if necessary, and find a community of friends and allies who share the same dream. With the insight of interbeing, we see that we're not a small me *realizing our dream, but a vast* us.

*And we arrive at our destination in every step.**

* Sister True Dedication's story is rewoven from a conversation we had, as well as other material. Hear more from her at climatewayfinding.earth.

Playlist

Songs to feel all the feels.

"A Hard Rain's A-Gonna Fall"—Laura Marling

"look up"—Joy Oladokun

"Resilient"—Rising Appalachia

Tune in at climatewayfinding.earth.

Deepen

Engage with your climate emotions through guided meditation, meeting them with compassion, curiosity, and even appreciation. (*~15 minutes*)

What you'll need: a quiet space to be in and the audio meditation, found at climatewayfinding.earth. (If you aren't able to go online and listen to the meditation, you can use the printed version here.)

You may find it especially fruitful to freewrite immediately afterward, using the journal prompts that follow.

Climate Emotions Meditation

If you'd like to offer this meditation to others, or ask someone to read it for you, here it is in print. Go slowly, allow pauses, and, if possible, model the breathwork.

Begin by finding a comfortable position to be in for the next few minutes. You may want to be seated; it might also feel nice to lie down. Whatever feels best for your body today. Take a moment to really get comfortable.

Once you're settled, allow your eyes to gently close, or soften your gaze, and shift your attention to the breath. Take a generous breath in, slowly and deeply—pause at the top of your full inhale. And exhale, slowly and long—with another pause at the bottom of your exhale.

Repeat that a few times. Inhale, pause. Exhale, pause. And now allow your breath to move into an easy, natural rhythm—in, and out.

As breath breathes you, notice the way your body is making contact. Feel the ground beneath you. Feel the firmness and steadiness of Earth—your bodily connection to the planet we call home. Notice the way the Earth holds you.

As you continue to breathe, bring into your awareness a place or a piece of the beyond-human world that you love. Maybe you know it well. Maybe you've only known it from afar.

What do you notice here? What colors, textures, light are present? Perhaps there are sounds, smells, or sensations on your skin. Behold what's before you and around you.

As poet Ross Gay says, rub a "sponge of gratitude" across it. Give thanks. Notice how that feels in your body, as you inhale and exhale.

As you breathe, let your awareness expand to the reality of the larger world in which we live—our climate and ecological crisis.

What images come into your mind? What sounds do you hear? Are there particular beings, communities, or places that come into focus? Who or what do you not see?

Notice what may be lost, and what is already being taken away. Notice these losses in the world, in your community, in your closest relations. *What do you see as you look to the future? And where do you not want to look?*

Bring awareness again to your body. *Where in your body do you feel the impact of the way we're going? How does it feel to hold the climate crisis in your head? In your heart? In your profoundly capable hands?*

Keep breathing—in, and out—as you explore the presence of climate emotions within you. Whatever emotions are arising, let them come. Whatever emotions are *not* arising, give grace to those too. Whatever you are sensing is exactly right for today.

Welcome and receive these emotions as a form of wisdom rising up from within you—you, a singular and connected node within the web of life. You are threaded into Earth's magnificent living systems. So you may be feeling Earth's pain. You may also be feeling Earth's power.

Each breath is a moment of connection. Connection to the life force that circuits between us and all beings across space and time. Inhale and receive that interconnection. Exhale and extend that interconnection.

Notice the steadiness of Earth, still present beneath you. Breathe in and feel how the Earth holds you. Breathe out and offer that holding to others.

Bring one hand, gently, to your heart and one to your lower belly. Take another slow, deep inhale, and another slow, long exhale.

As you are ready, slowly refocus or open your eyes.

(*If in a group*) Take a moment to take in this circle—shoulder to shoulder and present with each other's hearts—right here, right now.

Journal

Flow from your meditation experience into freewriting. If it's hard to identify specific feelings, you might look back at the Climate Emotions Wheel—or simply describe or sketch the experience of those unnamed emotions.

> What emotions arise in you around Earth's land, water, sky, and beings? How does it feel to hold the climate crisis in your head . . . your heart . . . your hands?

> What insights and guidance are your emotions offering?

Step Out

In the Zen tradition, questions can help us wake up. Sister True Dedication offers three such questions to consider while taking a slow, mindful walk outside: *Who am I? Where am I? What do I want?* (*20–30 minutes*)

+ Before you set out, you might find it useful to watch Sister True Dedication's explanation and guidance for this practice, found at climatewayfinding .earth.
+ As you walk for 10–15 minutes, focus on your breath. Feel the contact between your feet and the ground. Let the answers to each question open slowly, petal by petal. If you are physically unable to walk, move through space in whatever way works best for you.
+ When you've completed your mindful walk, find an object that speaks to your deepest motivations for climate engagement. Perhaps something symbolic, filled with memory, or evocative of particular emotions.

You might want to add this object to your "set apart" space for *Climate Wayfinding*. You might also share it, and the motivations it represents, with someone you trust. (If you're participating in a group, bring your object to the next gathering.)

Gather

This is a 75-90 minute experience.

Reminder: Everyone should bring their selected objects. (Consider putting a cloth in the center of the space where people may place these precious items.)

Tunes (*~10 minutes*)

Play songs from the playlist as your group gathers.

Poem (*~5 minutes*)

Read "Nothing Wants to Suffer" aloud. Then, invite a second reading.

Opening Circle (*15–30 minutes*)

This opening circle is unique and may take a good bit of time, depending on the size of your group. Offer some context first.

> In *Climate Wayfinding*, everything is "diver's choice." We go to whatever depth feels right for each of us. Engaging with our climate emotions, and doing so in community, may be appealing to some and off-putting to others.
>
> We want to bear in mind our range of lived experiences, which inform both how we feel about the climate crisis and the ways we might express or cope with those feelings. Family background, norms of gender, race, and culture, and personal encounters with climate impacts—it's all at play. Let's bring grace to ourselves and others, as there is no "right way" here.

Then, begin an opening circle with this prompt—modeling first, then popcorning. If you're gathered in person, place your object in the center of the circle after you share, and invite others to do the same.

> Share an object that represents or speaks to your deepest motivations for climate engagement. (If you didn't bring your object, describe it for us.)

Reflection (*~5 minutes*)

Ask the group to read through their freewriting from this chapter—and jot down any additional thoughts.

> What emotions arise in you around Earth's land, water, sky, and beings? How does it feel to hold the climate crisis in your head . . . your heart . . . your hands?
>
> What insights and guidance are your emotions offering?

Discussion in Trios (*~15 minutes*)

Invite everyone to form trios and share. Remind them to be in generous listening mode when they aren't the one sharing. Keep time (four 3-minute rounds) and let people know when to switch.

Discussion as a Group (*10–15 minutes*)

Return to a whole group, and invite people to share from their trios. If useful, bring in these prompts.

> Let's lean into the idea of emotions as guides. What came up there?
>
> Is there anything from this chapter that might invite us to see any of this differently or with more clarity?

Closing Circle (*~5 minutes*)

End with a closing circle using this prompt. Model it yourself; then popcorn.

> Sometimes our bodies can express things more effectively than our words. Share a body posture or movement that silently "speaks" to how you're feeling at the end of this session. When someone does their posture or movement, we'll all "echo" it with our own bodies.

Then, invite the group to take in this circle—shoulder to shoulder, together.

Closing Quote (*1–2 minutes*)

Read an inspiring passage from this chapter.

4

Surfacing Solutions

Up and out—look to the horizons of a (re)new(ed) world.

"I don't believe any longer that we can afford to say it is entirely out of our hands," James Baldwin said in the fall of 1960, addressing an audience at the *Esquire* symposium in San Francisco. He called their attention to "the paraphernalia of American life"—cars, refrigerators, the stuff of consumption—but more pointedly to what produces it, "and that is the person."

Baldwin charged that a country is "only as good . . . only as strong as the people who make it up and . . . turns into what the people in it want it to become."

In other words, society transforms not "by an act of God, but by all of us, by you and me." He rejected the idea that we do not have agency and placed accountability for the future in our hands: "We made the world we're living in, and we have to make it over."

Baldwin was not talking about the climate crisis. In 1960, the challenge of global warming was largely unknown to the public—though major fossil fuel companies were well on their way to understanding that all the excess carbon from their products would eject humanity from the not-too-hot, not-too-cold conditions of the last ten thousand years, basically just right for civilization.

But since I encountered Baldwin's counsel some years ago, I have come back to it again and again in the context of our planetary trouble. And I have held his words, sometimes with trembling hands, as an invitation to genuinely occupy our ability to make the world over in myriad, simultaneous ways.

As Earthlings of the twenty-first century, we find ourselves in an exceedingly rare and decidedly daunting position. We live and learn, labor and love, in a brief but decisive chapter of the human story—the chapter when we can still do *something* about climate change.

Even after decades of damnable denial and delay, there remains a slim window of time to move the world away from the worst climate scenarios and toward regeneration and resilience. Not perfection, not a return to planet past, not a future without gutting losses and real difficulty—but *better* is still available to us.

What we do, or don't do, in this decisive chapter will set the scene and the conditions for generations to come.

Some of us may feel we have hardly had any role in making our current world, especially as it relates to the climate crisis. Some may feel we have more of a role than we'd like to admit. But pointing fingers does not seem to be Baldwin's point. Whatever our starting position may be, he asks us to consider our part in what happens from here and step toward making this broken world anew.

> Pause on Baldwin's words: "We made the world we're living in, and we have to make it over." Perhaps say them aloud in your own voice. How does it feel to consider yourself a world (re)maker?

In making the world over, climate solutions will be key—all the ways and means we have to curb heat-trapping emissions and regenerate ecological integrity. (See "Sectors of Solutions.")

Solution is an imperfect word, suggesting something a bit too definitive, like landing all the correct answers to a crossword puzzle. But the root of the word—*solvere*—can help clarify what solutions are capable of and why they matter. *Solvere* is Latin for "loosen," and that is precisely what climate solutions can do. They can loosen the hold of current trajectories, incompatible with life as we know it, and weaken the grip of the fossil fuel industry on our global economy. With intention and care for how that loosening is done, especially as it affects marginalized communities, our cultivation of those solutions could also grow a more liberated, luminous world.

Today, our collective "toolbox" of solutions is robust and nearly overflowing—not deficient or pending, as some deceptively suggest and others unwittingly believe. There are so many practices and technologies already in hand and in motion.

I spent most of 2016 at my desk, writing most of the book *Drawdown*, which offered a groundbreaking, comprehensive look at the world's abundance of climate solutions. In essay after essay—from geothermal power to green roofs, peatland protection to plant-rich diets, silvopasture to solar hot water—the book's consistent, clarion through line is this: We have much of the knowhow and technical wherewithal we need to stop burning fossil fuels and come back into some balance with the planet's living systems.

There are even more measures we can take to speed the growth of those solutions and expand their reach—what I think of as *accelerators for change.*

And there are still other approaches to help communities cope and foster strength amid the challenges of mounting climate impacts—the realm of *adaptation* and *resilience.**

In all of this, I find welcome good news. There is much we can do and no reason to wait—*hallelujah!* That this is true may fill us with a sense of possibility and courage for the path ahead. Yet, in the midst of the good news, we may also find ourselves overwhelmed, even deflated.

Where to begin the remaking?
How to map the grand scheme of global opportunities to our bit of agency?
What tools to reach for first, to move society toward climate healing?

As author and activist Rebecca Solnit writes, "One of the best and most challenging things about the climate crisis is that there is no one solution. That is, the solution is a mosaic of many changes." This bounty is in some ways a gift, but also *a lot* for anyone trying to figure out where to concentrate and what to prioritize.

Given the scores of solutions, we may wonder what to pull from the toolbox—how to filter and focus, especially in our own climate engagement, since not a soul can do it all.

> Have you experienced solutions overwhelm? How does it show up in your body?

Quantification is the conventional approach to sussing out and honing in on solutions. Numbers have an important role to play. We do not have unlimited time or resources, so we want to know that we are focusing on interventions that

*All of these are *big* topics. If you are newer to them, you may feel that *Climate Wayfinding* barely scratches the surface—that's true. Going in-depth into climate solutions, accelerators, and adaptation strategies would make this an impossibly long book. Our focus will be framing those topics, and our relationship to them, in a way that supports sensemaking and next steps, including identifying areas to learn more beyond these pages. Find additional resources at climatewayfinding.earth.

can truly, meaningfully help. That's why you find cities, companies, communities, and campuses using greenhouse gas inventories and carbon footprint assessments. These methods of analysis point to problem areas—like fossil-fueled electricity, heating, and transportation—in need of solutions. Crunching data can yield useful clarity for beginning to answer: *What, then, shall we do?*

Drawdown is, in many ways, an expression of the quantification paradigm, as is the follow-up publication I led, *The Drawdown Review*. Each solution codified in those pages has a number next to it: the gigatons of carbon dioxide equivalent (CO_{2eq}) that a given solution could reduce or sequester between 2020 and 2050—either by stopping it from being emitted in the first place or by bringing it back down from the atmosphere.*

At a conference once, a fervent man approached me and launched into a full-throated recital of the more than one hundred *Drawdown* solutions. He named them in precise rank order, including their associated gigatons, all from memory. (He also had them printed on a T-shirt, dramatically revealed when, somewhat to my dismay, he unzipped and flung open his coat.)

The "solutions singer" was truly one of a kind, but fixation on the rankings the *Drawdown* research team had produced was widespread. *Numbers, numbers, numbers—the bigger the better.*

There was a lot about that numerical fixation that I could understand. It fell in lockstep with a common fondness for top-ten lists and the like—even though they radically oversimplify "best" and "most," placing everything from albums to restaurants on rungs of a stock standard ladder. I also understood the impulse to make the complex more comprehensible—to take the solutions overwhelm I myself had felt and soothe it with seeming simplicity. *Hey, just focus on the top five, the top ten.*

*CO_{2eq} is a measurement that puts all the different greenhouse gases, including methane and nitrous oxide, in apples-to-apples terms. A gigaton is one billion metric tons—roughly the combined weight of all livestock (mostly cattle) and all human beings on Earth.

Yet, amid the keenness for clean numbers and clear logic, I found myself saddened. In it, there was some inherent reduction of the incredible tapestry of solutions, which I had come to know intimately, and care quite a lot about, through writing those many essays. All the magic felt subsumed by the mechanical. All stats, no soul.

What's more, the rankings perspective obscured a critical insight: We need the whole system of climate solutions, and many of them are dependent on each other. For example, distributed solar photovoltaics benefit from energy storage, and linking rooftop solar panels and batteries together in microgrids is even better. Kelp farming in the ocean and improved rice production on land do not a regenerative food system make; they need dozens of other solutions growing alongside them. If we aspire for clean urban mobility, then public transit, bicycle infrastructure, and walkable cities are a powerful combo, producing a city you really can move around without a car.

Some years after the original *Drawdown* book came out, we decided to deemphasize solution rankings in future presentations of the work—in hopes of helping people see the forest, in its complexity and integrity, for the trees. When it comes to climate solutions, there is so much that is "necessary," "strategic," and "measurably impactful"—including the protection and restoration of *actual* forests.

In my experience, those credentials—"necessary," "strategic," "measurably impactful"—are insufficient for my own navigation through the breadth and depth of solutions. I run up against the limits of math as a decision-making, action-taking proposition, especially for the purposes of guiding my own deep climate engagement. The rational mind, I've discovered, can only carry me so far, and I need alternative or complementary approaches to hone in—and get fired up. I suspect most of us do.

To step, intentionally and actively, into the role of shaping our community, our country, and our world is no minor move. It feels less like a well-reasoned, matter-of-fact choice—something we do because it makes good sense—than it does a leap of faith—something we do because we cannot do otherwise. That's the kind of leap we can find ourselves taking when we make art, fall in love, adopt a pet, become parents, or feel connected to a higher power or the sacred.

> When have you taken a leap of faith–whether a long jump or a small hop? What stirred you to do so?

I think of artists, those who conceive and create the new from inspiration and seemingly thin air. A poet does not set out with a punch list so much as a stirring of something needing to find shape in words and the skill to craft it with imagery, rhythm, and meaning. A sculptor senses the form that their raw materials are calling forth and follows, chiseling intuition. Gardeners collaborate with soil and seed, pollinators and rain, to bring forth beauty and nourishment.

These are makers and remakers of worlds, small and large. What might be possible if we were to consider climate solutions in this spirit—not just as things to prescribe, plan, and push through, but as creative comrades to be in relationship with, remaking systems and society together?

In this decisive era, climate solutions need to grow everywhere and fast. All of them need human partners—friends, lovers, and playmates, so to speak. I believe our impassioned connection with climate solutions has an intangible yet potent role to play in that vital expansion.

When we orient to things that light a spark in us or tap a longing, we often find we can engage with gusto and stay engaged over time. Energy, I would posit, is vastly more powerful than information, catalyzing ingenuity and aspiration. In my own experience, that energy can be emotional, intellectual, spiritual, or some mix thereof, and I can feel it, physically, in my body. It might be a murmur; it might be a roar.

This approach, which we could term *solutions artistry*, doesn't mean that we never undertake humdrum projects or just do what needs to be done. (Some things, a friend says, are simply for the good of the order.) It certainly doesn't mean that our inspired efforts never founder or fall flat. Nor does it mean that we abandon rational and numerical understandings or toss all strategic plans.

It means that we also welcome knowing from some other place—the heart, the gut, perhaps the space right in between.

In *Climate Wayfinding*, we seek to give this heartful, gutsy knowing some structure and support, to help it surface and grow. We traverse the rich landscape of climate solutions, looking outward while tuning in to what stirs or emerges within us. As we do, we honor the relationships we already have, identifying what draws us to particular solutions and invigorating those commitments. We also sense for those we don't yet engage with but that pique our curiosity and generate a pull.

It's simpler than it may sound. Imagine looking at a delicious potluck spread or menu of dinner options. You might filter some out based on your dietary and nutritional needs. But, hopefully, that is just the beginning. Then, you intuit the ones you simply must taste, beyond those that seem perfectly fine. That is our energy and curiosity directing us, more rooted in aliveness and desire. For anyone who has lived through online dating, we have done something similar in that realm, too: swiping *left, left, left . . . right.*

When it comes to climate solutions, this approach can help us decide, again and again, where we want to learn more, where we want to dig in, and how deeply.

> What does *knowing from some other place* mean for you?

To grow the proverbial forest of solutions, accelerators for change are vital. (See "Accelerators for Change.") Solutions rarely speed or scale themselves—except where photosynthesis may be involved. Even then, we need ways to remove barriers they face, spur their adoption, and help them along. While accelerators don't have immediate effects on the pollution heating our planet, they can be powerful boosts for the solutions that do.*

The work of systems theorist Dr. Donella "Dana" Meadows helped unlock this for me. In 1999, near the end of her life, she published a seminal essay called "Leverage Points: Places to Intervene in a System." More specifically, her focus was on *complex systems*—the city of New Orleans, an ecosystem like the Miombo woodlands in Africa, the body of a narwhal, a corporation like Patagonia. "Leverage points," she writes, "are points of power."

In the essay, Dr. Meadows maps out twelve such points in increasing order of effectiveness. (Here the rankings speak!) They include negative feedback loops, such as when an electorate checks its representatives within the system of democracy, and rules, such as the policies those representatives pass into law. But far stronger, she argues, are the mindsets or *paradigms*—shared societal ideas—that produce a system itself. For instance, the idea that *forests are resources to consume* underpins industrial forestry, while the idea that *forests are sacred or ecologically essential* can catalyze efforts to safeguard them. We can transform those paradigms, or, even more powerfully, transcend them entirely.

I saw that in action when I spent time in Southeast Alaska, an archipelago of some 1,100 islands. While falling under the spell of old-growth forest and watery ribbons of inlets, I learned about the region's collaborative efforts to build a regenerative economy—one that exists in reciprocity with the Tongass. Dominated by towering Sitka spruce trees, the Tongass is the largest intact temperate rainforest in the world, a temple of biodiversity, and a vital carbon sink (somewhere nature stashes carbon).

In this home of bald eagles, brown bears, harbor seals, and many, many salmon, a well-tended network of leaders is breaking through the entrenchment of

*I developed an initial framework of accelerators for change when I was writing *The Drawdown Review*. Here, I've refined and expanded that thinking.

economy vs. environment—that long-standing, schismatic idea that asserts only one of the two can thrive. Step by step, they are proving that another paradigm is not only within reach but better for the forest *and* the people who live within those lands and waters. It's a fundamental retelling of the story, through both words and deeds. The implications of that retelling seared into me as I flew over the remains of colossal clear-cuts.*

To grasp this leverage point of paradigms, it can be useful to translate it into the accelerator of *changing narratives and culture*. The stories we tell, the beliefs we hold, the symbols we use to convey shared meaning, the rituals we participate in—these are all critical context for climate solutions and action. They tell us what's right or wrong, possible or impossible, necessary or not. Expressing, shaping, or reshaping societal ideas and shared zeitgeist is the very core of human communication and storytelling.

As you might guess, changing narratives and culture is the accelerator I find myself drawn to most fervently and most frequently—though for years I gave it short shrift, falling into that funny tendency to discount our own longings as somehow frivolous or less-than. Books, podcasts, films, op-eds, public talks, even social media posts—these are all units of cultural change. So are public murals, spoken word performances, sermons preached, and spectacles of all kinds. So are barstool chats and kitchen-table conversations.

> Can you recall a moment when something in culture—a book, film, work of art, public event, or the like—sparked a significant shift in your perception?

Reflecting on how the South African apartheid government was overturned, activist Dr. Kumi Naidoo puts a fine point on the power of culture: "We brought down apartheid with the stories we told, the songs we sang, the plays we staged. They controlled the means of violence, but we won the cultural struggle." And

*This trip was the basis for an episode of *A Matter of Degrees*. Find "The Tongass: A Way Forward for the Forest" at climatewayfinding.earth.

that opened the way for the first democratic election in the country's history, when South Africans selected Nelson Mandela to be their president.

Accelerators, though, are like solutions: None are singularly effective, and when it comes to healing our planet, we almost certainly need them all. We need to change culture and narratives; we also need to build power, rewrite the rules, shift capital, and more.

Over the years, I have come to imagine all the solutions, all the accelerators, all the strategies for adaptation and resilience more and more like a "wild rumpus," reminiscent of Maurice Sendak's fantastical scenes in *Where the Wild Things Are*. Or, for a more recent reference: *Everything Everywhere All at Once*, and everyone participating. We, artists of a (re)new(ed) world, are led by our energy and curiosity, longing and wonder.

It's a possibility that makes me feel buoyant, bordering on hope. But that buoyancy is not naive. Some things are called "solutions" but really aren't.* Many come with upsides and drawbacks both. So it matters significantly who is deciding, benefiting, or bearing whatever consequences there may be.

If we want to form partnerships with climate solutions, we must be willing to engage with their complexity. That means digging into the details, bearing in mind our own lenses and biases, understanding what might cause harm, and being willing to repair and pivot. We might find some to be too snarled for long-term relationships, or even a single date.

There are equally important debates around how to maximize what are often dubbed *co-benefits* of solutions—the beyond-climate good they can bring,

*In the book *I Want a Better Catastrophe*, Andrew Boyd proposes three questions for solution evaluation: "1. Is it a real solution? (Or fake greenwashing bullshit?) 2. Is it a just and democratic solution? (Or the same old inequality and exploitation, only now powered by renewables?) 3. Is it an on-time solution? (No matter how real or just it may be, is it simply arriving too late to make a difference?)"

from financial returns to public health. Researcher and educator Dr. Elizabeth "Beth" Sawin calls this *multisolving*, designing our interventions to advance not just climate protection but also food justice, transportation equity, biodiversity, good jobs, community wealth, and more.

These positive intersections may offer us another way into spirited connection with solutions, led perhaps by a commitment to gender equality, racial justice, rural livelihoods, or strengthening democracy. Just about anything we might care about—and anything amiss in the world we live in—connects with climate solutions and accelerators in some way. I imagine them each as poles that, together, can hoist a wonderfully wide tent—one spacious enough for our necessarily big, motley, rumpusing *we*.

> What do you already love or care about—from a casual hobby to a core conviction? How does climate converge with it?

Almost two decades after his *Esquire* address, James Baldwin was interviewed in *The New York Times*. Queried about literature and its entanglements with social change, he said:

> *The bottom line is this: You write in order to change the world, knowing perfectly well that you probably can't, but also knowing that literature is indispensable. . . . The world changes according to the way people see it, and if you alter, even by a millimeter, the way a person looks or people look at reality, then you can change it.*

Writing was Baldwin's way of responding to his own essential invitation to make the world over—to make it, as he penned in another essay, "a more human dwelling place."

So much about our world is out of your hands and mine, yet potential also sits right at our fingertips. We can unravel infrastructures and industries of harm and weave climate solutions so thoroughly into society that they become simply the way things are done. We can stitch the continuation of life such that it is not just viable but vibrant. How worthy, that handiwork.

Who might we become as we create the world—not tinkering from afar but wholly immersed in, and remade by, our relationship with this spinning sphere? And if we let our remaking be guided by land and water, shaped by sentience larger than ourselves, what beauty and brightness might suffuse this new, renewed world?

In the process, we might discover our true nature as makers, in loving, creative connection with the most extraordinary artist of all, this defiantly flowering Earth.

Sectors of Solutions

These solutions address key sources of climate pollution, largely by replacing coal, oil, and gas. They work in parallel with efforts to stop fossil fuel extraction and infrastructure.

Electricity

Generate clean, affordable electricity for all. Store and transmit it effectively. Use it as efficiently as possible to lighten the load. *For example*: wind power, pumped hydropower for storage, LED lighting.

Transport

Fuel mobility with clean energy and electric vehicles. Make alternatives to cars easy and effective. Increase fuel efficiency. *For example*: EVs, subways, walkable/bikeable cities, efficient ocean shipping.

Buildings

Electrify! Retrofit buildings to cut energy use and costs. Design new buildings to be clean and efficient from the get-go. *For example*: heat pumps to heat/cool, induction stoves, insulation, smart thermostats.

Industry

Make things with much less pollution. Put waste to good use with circular (not linear) flows of materials. Tackle refrigerants used for cooling. *For example*: alternative cement, recycled metals, composting.

These solutions curb pollution, aid carbon sinks, or both. Many are also key solutions for our nature and biodiversity crisis.

Food & Agriculture

Feed people well. Reduce food waste. Make diets more sustainable. Adopt farming practices that are better for land and water. *For example*: plant-rich diets, fertilizer management, agroforestry systems.

Land Ecosystems

Protect and restore terrestrial ecosystems. Bring degraded land back into beneficial use. *For example*: Indigenous peoples' forest tenure, grassland protection, abandoned farmland restoration.

Coastal & Ocean Ecosystems

Protect and restore aquatic ecosystems. Adopt better fishing/farming practices. *For example*: restoration of mangroves and salt marshes, seaweed farming, improved fisheries management.

Accelerators for Change

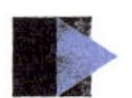

Change Narratives & Culture

Expressing and (re)shaping societal ideas through "units" of culture (e.g., stories, songs, symbols, performances, ceremonies).

Build Power

Fostering a strong, diverse *we* (e.g., through organizing, advocacy, campaigns, democratic processes, community building).

Rewrite the Rules

Using policy change and governance to facilitate more of what we want and curb what we don't (e.g., laws, taxes, incentives, rights).

Hold Responsible

Calling to account extractive, destructive entities to put a stop to their harm or seek damages for it (e.g., lawsuits, journalism, public pressure).

Shift Capital

Getting money out of climate harm (e.g., boycotts, divestment) and into equitable solutions (e.g., investment, grants, reparations).

Expand Knowledge

Building greater understanding of the problem, solutions, and their intersections (e.g., research, education, training).

Change Behavior

Shifting what's done and how, ranging from individuals to corporations and governments (e.g., shared goals, plans, norms, metrics).

Test, Learn & Try Again

Evolving beneficial technology and practices through conscientious, collaborative innovation (e.g., co-design, pilots, experimentation).

♥ Optimism

Jane Hirshfield

More and more I have come to admire resilience.
Not the simple resistance of a pillow, whose foam
returns over and over to the same shape, but the sinuous
tenacity of a tree: finding the light newly blocked on one side,
it turns in another. A blind intelligence, true.
But out of such persistence arose turtles, rivers,
mitochondria, figs—all this resinous, unretractable earth.

✡ Lighting the Way

with Valérie Courtois (Happy Valley–Goose Bay, Labrador, Canada)

Moving the whole of human society back into reciprocity with the land is a cardinal through line of climate solutions. It's a through line for Valérie "Val" Courtois, too, who deftly blends ancestry and strategy in her work on Indigenous-led conservation and stewardship. As she says often, "The majority of the world's remaining biodiversity is on lands cared for and loved by Indigenous peoples"—an unwavering true north.

Val is a member of the Innu Nation, a community on the shore of Peikuakami, or Lac Saint-Jean, in what's now known as Quebec. The Innu homelands lie at the eastern edge of the vast arc of boreal forest that covers more than half of Canada. It is the largest intact forest ecosystem on Earth—much of it still untouched by roads, cities, and industry—and an immense reservoir of life and carbon.

*My grandfather was a man who loved being out on the land. He loved being the early boat out to set casts and nets for landlocked salmon—*uananish *in our language—and I would go with him when I was able to visit. I have a distinct memory of when I was eleven. He was laughing as I struggled to pull up a net. I thought he was laughing at me, but he was laughing from the joy of seeing me in my element. "I hope you will always want to do this," he said. It was my job to be outside, to love and care for the land—and I've never forgotten that.*

At university, Val trained in the science of forestry. She was grateful to have professors who understood forests are too complex for Western science to explain. (Take photosynthesis, just to start.) Still, there was some unlearning to do when she began work as a professional forester and planner for First Nations.

When I was in school, forestry wasn't as focused on ecosystems; it was about cutting every stick. I was taught that my first question when I look at a forest is how to maximize economic return.

But as an Innu forester, my question is different: What needs to remain so that a forest can continue to be a forest, and I can continue to be Innu? We cannot hinder opportunity for future generations. That's a sacred responsibility, and it's why we have that instinct of saying what needs to stay *in the forest, as opposed to what can* go.

Val's approach to forestry honors art as much as science, mystery as much as management—all rooted in a love for place. As she became more deeply engaged in efforts to conserve the boreal forest, she saw the need for an Indigenous-led effort that would reach across Canada. It was time to center the unique relationship between First Nations and the land, for the benefit of the boreal, its peoples, and indeed all of society that depends on its ongoing vitality.

For Indigenous peoples, the land where we are from is at the heart of who we are. We are as much a part of the land and waters as the many animals and plants of this Earth.

For the Innu, Nitassinan*—which means "our land"—is a caribou landscape. We are a caribou people. Our understanding of our place in the world revolves around our relationship with the herds, such as the George River herd, just as our food security has historically. To continue to be who we are depends on large, intact landscapes.*

In 2013, Val worked with a number of other leaders to create the Indigenous Leadership Initiative (ILI); she is the founding executive director. The ILI focuses on stewarding boreal lands through the care of First Nations—a holistic strategy for self-determination, economic well-being, climate integrity, shielding biodiversity, and more. Their unique approach draws on Val's kaleidoscopic perspective, blending thousands of years of traditional knowledge, newly minted insights of science, and tools and structures of the modern world. It reaches to the past to shape the present and future.

The ILI's work ranges from lobbying governments to catalyzing funds to protect land. At its center is the work to uplift and promote a community of guardians, people who are watchers, researchers, protectors, educators, ambassadors, and keepers of culture. Many are engaged in managing Indigenous Protected Areas, places First Nations have identified for conservation, and in preventing the growing threat of wildfires.

Guardians help care for some of the healthiest, most vital lands on the continent. They ensure that our traditional laws are applied, that our communities are well informed of what's going on, that the nations have a strong voice in considering the pressures, options, and opportunities that come to them.

The benefits spread far and wide. As Gloria Enzo, a Ni Hat'Ni Dene guardian from the Northwest Territory, says, "We are sustaining our traditional territories not only for us, but for the whole world."

Val first learned about the model of guardians, and saw its magic, as a young forester in Labrador. It had been adopted there from a prototype on the Pacific Coast created by the Haida First Nation—the result of a battle with a logging giant.

When I showed up in Labrador from the city, I didn't know anything about that forest or how to survive in it. The guardians helped me really understand what my responsibilities were and how to do the job properly. I also saw their standing in the community. I thought, wow, if this could happen here, imagine if we had guardians across the country. Imagine what my own community could look like.

Val came to see guardian programs as offering not only a good career path but also a path to healing both Earth and humans.

As a person who's been a witness to and felt the intergenerational trauma of the colonial experience, I've found no better strategy for healing than nurturing our relationship with our place. The land heals, and I wish that for anyone who has experienced and lives with trauma.

But whoever you are and wherever you are, be connected to the environment you're in. It's such an important aspect of our humanity. Our elders say it all the time: If you want to heal, you go to the land. If you want to understand who you are, you go to the land. If you want to find your place in the world, you go to the land.

Yes, we live in a world of turmoil. But not all is lost. The land is resilient. We *are resilient. If we take care of the land, the land takes care of us.*

*This is about a relationship, a mutual love story.**

*Val's story is rewoven from a conversation we had, as well as other material. Hear more from her at climatewayfinding.earth.

Playlist

Songs for remaking worlds.

"Paradise"—Sturgill Simpson

"New World Coming"—Nina Simone

"This Was Made Here"—Meklit

Tune in at climatewayfinding.earth.

Map

Use creative cartography to further explore your relationship with climate solutions and accelerators for change. The prompts below will guide you to create a map—not in a traditional geographic sense but a visual representation of solutions you already engage with and those you might in the future. (*10–15 minutes*)

What you'll need: a two-page spread in your journal or a large sheet of paper; a pen, markers, and/or colored pencils.

A couple of caveats before you start:

+ *Content*: Enjoy this as a realm of imagining. If something pops into your mind, jot it down. Nothing is too small, and nothing is a commitment. When in doubt, include!
+ *Design*: Use any organization (or lack thereof) that feels in keeping with how you experience the world. The aesthetic bar is low—go wild with words and doodles.

If you'd like to see example maps for inspiration, find some at climatewayfinding .earth.

These prompts will take you, step-by-step, through populating your map of solutions and accelerators. As something pops into your mind, jot it down. (You might want to play music while you map.)

Put your name in the center of the page.

Begin by mapping relationships you already have with solutions and accelerators. Look close-in—at your home and day-to-day life.

What do you eat? How do you move and travel?
What do you wear? Purchase? Choose not to purchase?
Where do you bank or invest savings? (Yes, there are climate-friendly options!)
What do you read about? Talk about with friends? Post about online?

Add anything close-in that needs a place on your map.

Now, continue to map relationships you already have with solutions, but farther out—in your neighborhood, your town or city, your workplace, or beyond.

What are you learning or researching? What classes are you taking?
Have you had climate-related jobs? Perhaps you have one now?
What groups are you part of? Any climate-related projects or campaigns?
How do you engage in the broader community and civic life?

Add anything farther-out that needs a place on your map.

Now consider *what could be*—relationships you might form with solutions and accelerators in the future.

Which ones are sparking your energy and curiosity? Which feel most vibrant?
What do you want to learn more about? What do you want to try? Or take part in?
Are there things you'd like to do more of? Or less of?
What beyond-climate needs or opportunities call to you?

Add anything prospective or aspirational that needs a place on your map. Blue-sky and zany ideas are most welcome.

Take another moment to scan your landscape of solutions and accelerators—whether drawn from the past, happening in the present, or possible in the future. Make any last additions.

When your map feels complete, shift to the journal prompts below. (If you're participating in a group, bring your map to the next gathering.)

Journal

Take a moment to look back over your map of solutions and accelerators. Then, begin to freewrite.

> Sense into your energy and curiosity, your longing and wonder. If you were to follow their lead, what solutions and accelerators would you focus on? (Maybe circle or star them on your map.)

> In what ways are you already an artist of a (re)new(ed) world? Releasing all practicalities for now, in what ways do you wish to be?

Step Out

Sometimes learning more is the best next step to take. That's especially true for the vast landscape of climate solutions and accelerators for change. *(20–30 minutes)*

+ First, select one solution or accelerator that is especially alive for you right now, that you'd like to grow a relationship with. Quickly brainstorm the questions you're holding about it.
+ Then, go foraging for answers, and notice what stirs in you as you do. You might search for examples of that solution or accelerator in action. You might find an expert or group working on it, perhaps locally, and dive into their work. You might read an article or queue up a podcast episode. (If you're prone to rabbit holes, setting a 25-minute "Pomodoro timer" can be useful.)
+ Finally, close your eyes and imagine what your relationship with this solution or accelerator might look like going forward. Sketch images or jot notes of what you envision.

You'll find some resources for research at climatewayfinding.earth. 

Gather

This is a 75-90 minute experience.

Reminder: Everyone should bring their solutions and accelerators maps. (Consider bringing materials for pinning up and interacting with everyone's maps—e.g., wall-safe tape, sticky notes, markers.)

Tunes (*~10 minutes*)

Play songs from the playlist as your group gathers.

Poem (*~5 minutes*)

Read "Optimism" aloud. Then, invite a second reading.

Opening Circle (*5–10 minutes*)

Begin an opening circle with this prompt—modeling first, then popcorning. (Bring a particular spirit of brevity to this one, as it's a full agenda today.)

> Share one thing (climate-related or not) that has sparked your energy since we last gathered.

Group Agreements (*~5 minutes*)

Bring everyone's attention back to the Group Agreements (from the first chapter). Invite discussion on how they're going and whether any adjustments are needed.

> As we approach the halfway point of the journey, it's a good time to check back in on the Group Agreements. Let's take a minute to read through them. How are we doing with the edges of our collective garden? Are there any that need tuning?

Map Browsing (*10–20 minutes*)

A little something different for this gathering: Ask everyone to pull out their maps and put them somewhere visible. (They can be placed on tables or taped on the wall. Or they can be photographed and shared, if the group meets online.)

Then, invite the group to browse, admire, and engage with each other's maps. (Sticky notes work well in person. If online, a comment or reply feature works nicely. You might play instrumental music during this activity.)

> It's immensely powerful to tap into the group's collective wisdom and support. As we read through the maps, we'll leave words of appreciation or encouragement, offer useful resources, or express interest in connecting.

Discussion in Trios (*~15 minutes*)

Invite everyone to form trios and share from their experience of making a map and exploring others'. Keep time (four 3-minute rounds) and let people know when to switch.

Discussion as a Group (*~10 minutes*)

Return to a whole group, and invite people to share from their trios. If useful, bring in these prompts.

> Relationships, artistry, creative partnership—how does it feel to think about climate solutions in these terms? Does anything shift in how you might engage with them?
>
> This chapter covered a lot of ground. What is tugging at you for further exploration?

Closing Circle (*5–10 minutes*)

End with a closing circle using this prompt. Invite someone to start; then popcorn.

> Share one (yes, just one!) climate solution or accelerator you're feeling especially jazzed about.

Closing Quote (*1–2 minutes*)

Read an inspiring passage from this chapter.

5

Offering Our Superpowers

Turn inward, again. Find the fields of your bent, your brilliance.

I live in the very heart of Atlanta, Georgia, affectionately called the "city in a forest."* From my desk, where I work most days, I look out onto a stand of trees. Right at canopy height, it is the perfect view for getting distracted, especially by our resident red-tailed hawk, who is strikingly visible in the loose thatching of bare winter limbs.

Sudden squirrel scatter, and she alights on the branch of a maple tree to scan for potential prey. Her fleet perch and keen watch, her grandeur of feather and hunt—it breaks through the primacy of my screen and shakes me from the fathomless digital world. Interruption gladly received.

Each time the hawk stops through these trees, I am struck by the sudden proximity of a taloned huntress to me, encased in my condo-version of captivity. More than once, I have grabbed my phone to quickly frame the hawk and catch ill-focused evidence that I too am alert and alive. Enraptured by a raptor, I have "Slacked" the flattened scene to my colleagues: *Afternoon visitor!* 🪶🖤 (As if icons in miniature could limn her.)

But I am struck by another proximity, too, between what the hawk does and who the hawk seems to be. Her *doingness* and her *beingness* are so close as to become one.

I suspect this hawk has never once felt the nag of the question, *What can I do?* Not about the climate crisis. Perhaps not about anything. What to do is something other animals seem to know innately and intimately, or perhaps don't need to *know* at all.

Evolution has made things more complicated for us *Homo sapiens*, who ponder and puzzle. As essayist and author Margaret Renkl writes, "Every living thing—every bird and mammal and reptile and amphibian, every tree and shrub and flower and moss—is pursuing its own vital purpose, a purpose that sets my human concerns in a larger context." As I watch the hawk's wings lift and lower and propel her back into the air, I marvel and muse whether life itself might offer another way in.

What might open up for us if we shift the question ever so slightly—from *What can I do?* to *Who can I be?* Or, *Who am I already?*

*In an era of climate crisis, urban tree canopy—the constellation of leaves and branches that cover a city when viewed from above—is critical infrastructure. Through shading and transpiration, trees reduce the effects of soaring heat and ease the impacts of increased rainfall. As I wrote for *CNN*, being the "city in a forest" is more than beauty and good branding; protecting, as well as replanting and expanding, Atlanta's urban tree canopy is a matter of safety, equity, and resilience.

The hawk, like all of us existing on this planet, is an inheritor of a 3.8-billion-year history: From single-celled organisms to plants and vertebrates, life has continued to move forward toward more life, overcoming unthinkable odds. Weighty and unwavering and in so many ways impenetrable—this dynamic defines Earth as a *living* planet. When we think about a hive of honeybees gathering their ingredients from flowers, or black corals siphoning plankton over centuries, or the sudden emergence of mushrooms from a shrouded fungal network, we can see this dynamic in action. Even kudzu offers testimony with its rampant return, however unwelcome, each spring.

Who can we be? One thing we already are: an expression of Earth's life force, right here, right now, made possible by a series of miracles that have blossomed over eons. This is true simply by virtue of breathing.

Life force unfurls through each of us in such beautifully different ways. We explore the unknown and document our discoveries. We design new things and give them form. We expose what's ruptured and source the means to mend it. We reflect, wonder, and imagine. We craft stories and art and shows. We make ritual. We convene people and foster conversation and collaboration. We care for one another. We strategize, organize, and orchestrate. We engineer and implement. We manage the details. We show up, stand up, and speak up. We share wisdom and tell jokes. We cook and sing and clean and plant and build and nap. And all of that is just the briefest inventory of human beings' doings.

There are things we do that are so wholly connected with who we are—that spring up from within us in such an organic way—that the space between our doing and our being shrinks or even vanishes. In those moments, our small expression of the vast life force we've inherited and embody is especially effervescent. We may find ourselves buzzing, flowing, or sensing a particular warmth. We may be especially porous and focused both.

I imagine this is how the hawk might feel as she swoops into the circle of life. It's how I wish many more of us to feel as we take wing to heal the climate crisis.

> Take a moment to look out your nearest window—or around you, if you're outside. Where do you see aliveness manifest or in motion? Let it seep in.

In *Climate Wayfinding*, we think of the ways we each express life force as our unique talents, gifts, or *superpowers*, all of which are so very needed in this era of change. (Please note: superpowers in the DC or Marvel sense, not world-dominating countries.)

Sometimes, the ways we shine are eminently clear. But other times, our own strengths can have a way of hiding from us. We may exhibit them but struggle to name them. That's especially true if they are not rewarded in school or by our economy and culture, or if they are sundry and a bit hard to pin down. Singular, exceptional talents may be more insistent as to how we should spend our days. *Be a jazz singer or a goalkeeper, of course!*

Clarity may also be muddled if we think our knowledge and skills are not so relevant to the context or topic at hand—in this case, our planetary troubles. *Climate Wayfinding* takes as a central conviction that in our time of transformation, we need a vast array of talents. We need whatever you and I have to give.

So, we bring a paired set of lenses to help illuminate our unique gifts: *authentic power* and *deep joy*.

The word *power* carries many layers of meaning, not all of them good. It might call to mind the idea of power over others—hierarchical authority or control in politics, business, or broader society, which when gained by some often results in a loss for others. It might evoke physical strength or force exerted on someone. Or it might chime in your ear as a key synonym for electricity—*Oy, there's a power outage.*

But *authentic power* is something different. It's power that rises up from within us—internal and genuine, not gained at others' expense or expended

The Nexus of Power and Joy

upon them. Authentic power is a feeling of ability, capacity, strength, weight, energy, vigor. It aligns what swells within us with how we move in the world. (We will turn to notions of collective power, beyond the individual, in the next chapter.)

In the wake of a cancer diagnosis at age fifty, Audre Lorde, the writer-philosopher we met earlier, wrote a radiant essay on living with cancer. She conveys her own authentic power in blazing terms:

> *No matter how sick I feel, I'm still afire with a need to do something for my living. . . . I want to live the rest of my life, however long or short, with as much sweetness as I can decently manage, loving all the people I love, and doing as much as I can of the work I still have to do. I am going to write fire until it comes out my ears, my eyes, my noseholes—everywhere. Until it's every breath I breathe. I'm going to go out like a fucking meteor!*

Lorde also points in the direction of the other vital lens we use in *Climate Wayfinding*, which is *deep joy*. It's a feeling of great pleasure, happiness, delight, exhilaration, radiance, bliss. Deep joy, too, rises from within and spills out, intermingling with the world around us. It is often the emotional glow of meaning or connection—in the Greek sense of *eudaimonic happiness*, growing not so much from sensation but from significance.

For some, joy may be a less complicated notion than power. Joy feels, perhaps, more infinite or more beneficial for all; at the very least, benign. Yet it may also feel out of place in the face of the climate crisis. Who are we to taste joy when so much is hurting? To rejoice when so much is being lost?

But joy is all the more necessary, and all the more holy, in difficult times. In *Braiding Sweetgrass*, Potawatomi botanist and author Dr. Robin Wall Kimmerer writes:

> *Even a wounded world is feeding us. Even a wounded world holds us, giving us moments of wonder and joy. I choose joy over despair. Not because I have my head in the sand, but because joy is what the Earth gives me daily and I must return the gift.*

Rather than deep joy being an act of turning away from life, it's actually a way to turn toward, tune in, and take part. As Mary Oliver counsels about joy: "Perhaps this / is [life's] way of fighting back, that sometimes / something happens better than all the riches / or power in the world." And when it does? "Give in to it," she urges. "Joy is not made to be a crumb."

For joy as well as power, it needn't be all sweetness and light—both can emerge amid and even from adversity. Our great gifts can have a way of growing from things that scrape and gash and scar. Even in the shadow, there is a glimmer to be found.

> Who or what embodies power and joy for you? What does that look, sound, and feel like?

In my own experience, moving at the nexus of authentic power and deep joy might be our closest approximation to life force itself. There are things that land right in the sweet spot between them, giving a sense of power rising up and joy spilling over all at once.

For me, that nexus includes shared exploration of ideas and possibilities. Feeling the world around me and inside me. Weaving words. Putting pieces together to manifest something new. Witnessing people in their unvarnished emotions, wisdom, and magic. Intimate, well-held circles and classrooms. Poets poeting and songwriters in song. Creative partnership. Connection with land and animals. Immersing my body in water. Holding trees in awe. Moon, too, as she comes and goes. Being with old friends or discovering new ones. Witty banter. Catharsis and growth. Giving voice to who I am or what I hold precious and true. Moving with thousands of others in a march. The embrace of someone I love.

To be sure, not all of it has direct relevance to my climate engagement. Much of it does, though, which has been increasingly the case as my career has evolved and I have had more opportunity, and gumption, to make that so.

When I have strayed far into zones of *not*-power and *not*-joy—most often for employment or another hard-tugging *should*—I have found myself in struggle, disconnection, and even depression. Over the years, I have gotten better at not only reading the melancholic revolt of my soul but responding to it compassionately and actively—like making my way, when possible, to a different job, one that gives me more chances to give of the best of myself. Stubborn is the soul, intent on a space where it belongs.

I have learned, too, that tending to the power-joys that are utterly unrelated to my climate engagement—like caring for and training a quirky, darling horse named Birdie—is not actually unrelated to my climate engagement at all. They are part of what keeps me committed to this whole thing we call life, the continuation of which is the very point of planetary healing.

I cherish a framed photo of myself, not even two years old, staring with glee at a palomino pony on the other side of a wooden fence. Crouched in my tiny pink overalls, eager fingers on a fence board, I look like I am ready to spring forward with every ounce of my being. It reminds me of the things our bodies tell us before we can even put them into words: We are made for elation and connection and love, and that is what makes life so wholly worthwhile.

After years of trying to fill a horse-shaped hole in my adult life with something other than a horse, I am finally honoring that embodied truth. I rarely feel more alive than I do with those creatures. And after an early morning at the barn, I return to climate work with higher spirits—more equipped to participate in the exchange of energy that makes Earth flourish.

Given the uphill efforts ahead of us, having ways to renew and reinvigorate is so very important. And we may *want* some such things to be set apart, even safeguarded, from the ins and outs, ups and downs of a climate journey. For me, that kind of sanctuary is essential.

> What do you turn to for soul-filling renewal? And where do you find sanctuary for your life force?

It is a radical act to believe in our ability to thrive, both individually and as a planet, by being who we are. I mean *radical* in the fullest sense: from the root, fundamental, and far-reaching. A person anchored and aglow—that is the kind of revolutionary that's called for in this time.

Yet expressing our power and joy, generously giving our gifts to the world—these may feel a bit idealistic or foolish in the face of important practical considerations, such as steady employment, student loans, housing, ailments, obligations to family or other loved ones. How do we manage it all at once?

Certainly, not all necessary climate work is paid. Setting up neighborhood compost pickup, attending a utility commission hearing, running a divestment campaign—these may not be how we keep a roof overhead and food in the pantry. But, happily, more and more we *can* make a living by helping to build a climate-safe and just future. Some jobs are made that way. When that's not the case, people may find ways to refashion their work with Earth in mind. As advocate and coalition-builder Jamie Beck Alexander has argued, very nearly "every job is a climate job," or can be.

Indeed, ridding our economy of pollution, fostering community resilience, and restoring healthy ecosystems will require millions upon millions of doers—and could very well create millions upon millions of new good jobs. We will need many more initiatives like the American Climate Corps, a national service program maddeningly axed in early 2025, to bring young people into the work and equip them for meaningful, planet-mending, financially sound livelihoods.

Creating the future we want will require curious *observers*, detailed *analysts*, and ecocentric *naturalists*; galvanizing *organizers*, cultivating *growers*, and get-it-done *builders*; illuminating *communicators* and relational *careworkers*. Mapped out by a Sunrise Movement initiative called Green New Careers, each of these professional archetypes will be indispensable, and all will be interdependent. (See "Climate Career Types.")

Consider the progress we have made on climate to date. It would not have been possible without astute researchers and land stewards clueing us into the problem and its causes. Visionaries helping us see other paths. Inventors shaping solutions and technicians getting them up and running. Advocates pushing for change in a system that resists it. Lawmakers shepherding needed regulation. Entrepreneurs trying to recast industry. Frontline responders helping communities get through disaster. And much more. Many are examples of how to turn the work of climate healing into a vocation we can sustain.

Any one role can seem small in isolation—like any single species in the biological community of an ecosystem. Just a wee skittish salamander. Just a spindly spider weaving a fine web. But it may be that very smallness—particularity, uniqueness, and niche-filling—that facilitates the genius of the bigger whole. Our collective talents engender tomorrow in very real ways. So, today, we need many more of us discovering, deepening, and evolving *who we can be* in this awesomely diverse, increasingly formidable ecosystem of transformation.

> Is there a gift you feel called to give? A generosity you might offer, simply by being who you are?

Looking inward to shape our outward contributions—this, I think, is a form of courage. When we refuse to lose touch with our sources of authentic power and deep joy, and when we dare to center them somehow in our lives, we reach toward calling. Whether loudly or in a whisper, these things summon us, insisting that our lives can be *alive*—sprouting and blooming, swooping and flying high—and that we can be part of making it so.

When we hew to power and joy in how we take part in climate healing, we have an opportunity to infuse the whole endeavor with a different kind of spirit. We can foster an ethos of potency and jubilation, rather than a dour, finger-wagging, let-me-fact-you vibe. Without adopting toxic positivity or bypassing the severity of our crises, we can let goodness roll and romp.

All the things that light us up could help the climate community—and the collective effort of what Joanna Macy called the *Great Turning*—glisten like a party people might actually want to attend.

Just now, the red-tailed hawk draws my eye. It's a beautiful, bewitching thing to behold a being in the fullness of herself. But I realize, watching her in motion, that I am rapt by more than the solitary bird. At the edges of the self, there is a zone, almost an aura, of arising. We find there, at the periphery, a space populated by all that is emergent with, and only with, the world around us.

For the hawk, that emergent edge exists in the remarkable everyday interplay of hunger and wing and wind. For Audre Lorde, it was where her fine-toothed attention and the act of composition would meet a reader turning the page. And perhaps it is so for all of us, along our own edges, as we muster skill and strength for a planet in want, in wish.

Perhaps you, too, can feel the vibration at the eager verge of *doingness* and *beingness* and the wide, long, insistent breath of life.

Climate Career Types

These archetypes are adapted from Green New Careers, an initiative designed by the Sunrise Movement to help people see their potential roles in the future we could build together. What feels most resonant to you?

The Careworker

A people person, who helps individuals, relationships, and communities move toward health and well-being. *Kind, generous, empathetic, patient, cooperative.*

The Observer

A curious being, who delves into complexity to probe, understand, and share how our world works. *Intellectual, inquisitive, methodical, analytical, rational.*

The Naturalist

An ecological creature, alert to the land, water, and air, who works to repair the world's natural rhythms. *Clever, systematic, passionate, effective, observant.*

The Communicator

An expressive creator, who helps us see what is happening today and imagine what could be tomorrow. *Creative, intuitive, sensitive, articulate, artistic.*

The Organizer

An adept wrangler of others, who brings people together across differences to work toward shared goals. *Assertive, enthusiastic, confident, persuasive, motivational.*

The Builder

A get-it-done person, who sees the need at hand, grabs the tools required, and takes up the task willingly. *Efficient, genuine, persistent, practical, handy.*

The Grower

A cultivator, who seeks to share the Earth's bounty—food, water, medicine, materials—while keeping balance. *Measured, hardworking, flexible, sturdy, trustworthy.*

The Analyst

A detail-oriented person, who uses data and analysis to guide the decisions we make and the directions we take. *Conscientious, organized, precise, meticulous, logical.*

An Old Story

Tracy K. Smith

We were made to understand it would be
Terrible. Every small want, every niggling urge,
Every hate swollen to a kind of epic wind.

Livid, the land, and ravaged, like a rageful
Dream. The worst in us having taken over
And broken the rest utterly down.

A long age
Passed. When at last we knew how little
Would survive us—how little we had mended

Or built that was not now lost—something
Large and old awoke. And then our singing
Brought on a different manner of weather.

Then animals long believed gone crept down
From trees. We took new stock of one another.
We wept to be reminded of such color.

✡ Lighting the Way

with Wahleah Johns and Billy Parish (Berkeley, California; Tónizhóní, Navajo Nation)

When we orient our lives to climate healing, we set ourselves on an ever-evolving path, not a linear one. With a sustaining sense of purpose, we can respond to shifting circumstances, and our climate engagement can find new forms across the seasons of life. For clean-energy visionaries Wahleah Johns and Billy Parish, their distinct yet braided paths show how we shine in different ways at different times—and how the heart of the work carries us through.

A member of the Navajo (Diné) Nation, Wahleah is from the high desert of Arizona, atop the sacred female mountain of Black Mesa. She grew up with a keen awareness of the fossil fuel economy and the ways government and industry connive to strip communities of their rights so that land can be stripped of its resources—in this case, coal and water.

For decades, Peabody Energy operated two mines there, sending much of that coal via a slurry pipeline to a power plant in Nevada. To make that semiliquid slurry, Peabody pulled more than a billion gallons of pristine groundwater from the Navajo aquifer every year—water needed for people, farming, and ranching, which had been Wahleah's grandmother's livelihood.

This is our only groundwater source in a region that gets less than seven inches of rain per year. The traditional people who do offerings to our springs noticed that they were starting to dry. It caused a lot of concern, and it set me on my journey. Navajo and Hopi youth started mobilizing to study Peabody's impacts and raise awareness with tribal leaders. We formed the Black Mesa Water Coalition in 2001.

Wahleah could see clearly how the Navajo were being ripped off, barely paid for their precious water. She went full bore into community education and organizing—and the thorny politics that surrounded coal.

In 2006, that work took her to New York for a strategy meeting held by the Energy Action Coalition, a national alliance of youth climate organizations that Billy Parish was running at the time.

Billy grew up in New York, on the Upper West Side of Manhattan. In high school, reading Daniel Quinn's book *Ishmael* woke him up to the ecological cliff humanity has created, and it left him grappling with how best to help. In college, that intellectual awakening became profoundly visceral during a research trip to the glacial source of the Ganges River, deep in the Himalayas.

I saw climate change face-to-face and how catastrophic the impacts were going to be. The source of drinking water for hundreds of millions of people was receding every year. And that wasn't just the case for the Ganges but for major rivers throughout Asia. I collapsed onto the ground in tears. That was my moment of no turning back: This is what I'm going to spend my life working to mend.

Billy became passionate about creating on-ramps for people to engage in solutions to this enormous challenge. When we met as Udall Scholars in 2003, he was already on the verge of leaving academia behind. He took a semester off from undergrad to focus on building a movement of young people across the US. Then, he took another semester off, and then another.

By the time that strategy meeting took place in New York, the Energy Action Coalition was many campaigns in. Climate became somewhat secondary for Billy as the day unfolded and he got to know Wahleah.

She was very quiet at first but listened intently. And when she spoke, everybody in the room paid attention—everything she said was meaningful. Then, at a fundraiser that evening, she gave a prayer in her language. I was struck by how grounded she was. Afterward, a bunch of us went to play basketball, and there was a moment where she crossed this guy up, took him to the hole, hit this floater—damn. By the end of that, it was over. I was smitten.

A movement meeting ultimately led to marriage in 2007. Billy and Wahleah made a home together in Arizona and started a family. It was a crossroads for both of them, personally and professionally.

The Black Mesa mine had closed in 2005, as coal had become a more costly way to produce electricity. But Peabody's other mine continued to operate with impunity despite vigorous organizing. Wahleah found herself at odds with her neighbors who wanted mining jobs to continue.

Saying no to coal was a big deal in my nation, where I was labeled radical by my own people. There was lots of tension, arguments. It was really hard.

When I became a mom, I got quiet. It put me in a really gentle space and made me think about the future. What are we standing for—for our children and their children and their children? What's the seed we can plant that Peabody cannot touch? I started to go deeper into meditation and prayer around that.

Where her sights had once been set on fighting Peabody, Wahleah was now looking beyond coal. More than resisting fossil fuels, she wanted to foster renewable energy for her community—and ensure a fair and just transition from one energy system to the next.

I got really excited about building solar for the Navajo Nation and what we could do together. I knew how to organize communities, so we started those conversations, that visioning—not thinking about today or yesterday but thinking about the future. I loved that work.

That work led to the founding of a solar-installation startup, Native Renewables. There was simply no reason for eighteen thousand Navajo homes to lack access to electricity when they could tap into the power of the sun.

Billy's attention was also shifting toward the good we need to build. He realized business would have to play a role in bringing climate solutions to scale and could create jobs for the many young people who want careers in climate. As with Wahleah, Billy, too, found that a new family inspired new priorities.

I had burned myself out running the Energy Action Coalition. I felt like I was carrying the weight of the world on my shoulders. I needed to restructure my life from 100-hour workweeks, constantly on the road. Time became more precious, and we needed to make money to care for our kids.

Billy's get-everyone-involved ethos found a new expression in a solar-funding company, Mosaic. After some fits and starts, Mosaic also came to focus on residential solar, pairing better loans for homeowners with a better experience applying for them. It brought solar lending into the mainstream—to the tune of almost $20 billion in loans to make electricity right on our roofs or in our backyards. Though he was building a tech company, widespread participation remained the core of Billy's purpose.

How do we come together to build something more beautiful? I've always wanted to create containers to enable that. Make it easy, affordable, compelling. And for me, the feeling of working with an amazing group of people on something that really matters—that's just where I want to be. That's my flow state.

In 2021, Wahleah had an opportunity to dramatically expand the reach of her work. She became the director of the Office of Indian Energy, part of the US federal government. For four years, she worked across all 574 recognized tribes to expand access to solar and wind power.

Tribes are leading the way in clean energy, and tribes are resilient. We have a history with this country that goes way back, and we have remained connected to who we are, to our land and identity. Even when the federal government isn't supporting us, we will keep going. The journey of renewables is going to continue. The power is in our hands.

Now, Wahleah and Billy are at another crossroads. Her tenure in government has ended. He has handed the reins of Mosaic to a new CEO. They've both been making time for rest, for dreaming, and for returning to Black Mesa, which Wahleah describes as "a source of its own natural healing." Billy says, "It's like recharging a battery." He's lit up, too, by what's happening beyond the Navajo Nation and beyond the US.

Clean energy has gotten so cheap—it's taking off all around the world. In Pakistan, people are buying solar panels at farmers markets, then watching TikTok videos to see how to connect them. When I started this work, the technology wasn't there. We could only make a case based on altruism. Not anymore. That feels really, really good.

But when the road is tough—and it can be—it helps me to remember that this is our life's work. The only way is through.

When something becomes our life's work, it shapes more than our workdays. As with Billy and Wahleah, it may lead us to love and to land and to family. It asks us to grow into more aligned and generous versions of ourselves. As we do, our light can shine brighter and truer, like a small, steady sun.*

* Wahleah and Billy's story is rewoven from a conversation we had, as well as other material. Hear more from them at climatewayfinding.earth.

Playlist

Songs to celebrate our beingness.

"Bird Song"—The Wailin' Jennys

"I Am"—Beautiful Chorus and India.Arie

"Never One Thing"—May Erlewine

Tune in at climatewayfinding.earth.

Deepen + Map

Intermingle guided meditation and creative cartography to uncover, explore, and map your sources of power and joy. (*~15 minutes*)

What you'll need: a two-page spread in your journal or a large sheet of paper; a pen, markers, and/or colored pencils; the audio meditation, found at climatewayfinding.earth. (If you aren't able to go online and listen to the meditation, you can use the printed version here.)

+ Set up your power and joy map: on one side write *power* and on the other side write *joy*.
+ Begin the meditation, which prompts you to reflect and then map at two key junctures.
+ When your map feels complete, take up the journal prompts that follow.

If you'd like to see example maps for inspiration, find some at climatewayfinding .earth.

Power and Joy Meditation

If you'd like to offer this meditation to others, or ask someone to read it for you, here it is in print. Go slowly, allow pauses, and, if possible, model the breathwork.

Begin by coming into a comfortable position for your body. Maybe you're sitting up. Perhaps you're lying down.

As you settle in, allow your eyes to gently close or soften your gaze. And notice the firmness of Earth beneath you. Begin with a generous breath: breathing in . . . and out . . .

You may want to put one hand on your heart and one on your lower belly—connecting to these two wisdom centers of the body. And continue to breathe.

Imagine your body full of a sense of genuine, authentic power. *Power*—ability, capacity, strength, energy, weight.

> *How does power feel in your body?*
> *What sensations do you notice?*
> *How does it shape you, physically?*

Now, imagine yourself doing, experiencing, or participating in something that brings you a sense of authentic, genuine power. Feel yourself in that activity. *What do you notice?*

Allow another thing that brings you a sense of authentic power to come into your mind . . . and now, allow another . . .

Staying anchored in your body, open or refocus your eyes, write these first things down, and then allow a freewrite stream of things that bring you a sense of authentic power. Take a few minutes to write down whatever arises.

(*Pause for 3–4 minutes.*)

Come back to a comfortable position for your body. Allow your eyes to gently close or soften your gaze a second time. And tune in again to the firmness of Earth.

Return to a generous breath: breathing in . . . and out . . .

You may want to reconnect, hand to heart and hand to lower belly. And continue to breathe.

Imagine your body full of a sense of deep, resonant joy. *Joy*—great pleasure and happiness, delight, exhilaration, bliss.

How does joy feel in your body?
What sensations do you notice?
How does it shape you, physically?

Now, imagine yourself doing, experiencing, or participating in something that brings you a sense of deep joy. Feel yourself in that activity. *What do you notice?*

Allow another thing that brings you a sense of joy to come into your mind . . . and now, allow another . . .

Staying anchored in your body, open your eyes, write these first things down, and then allow a freewrite stream of things that bring you a sense of joy. Again, take a few minutes to write down whatever arises.

(*Pause for 3–4 minutes. Then, flow into the journal prompts below.*)

Journal

After you complete your map of power and joy, take a few minutes to read through what you've captured. Admire! Appreciate! Then, reflect and freewrite.

- What did you uncover about authentic power? What did you uncover about deep joy? If they overlap, is there significance in where the two meet?
- What stands out on your map? What's here that you want to offer our world or center in your climate contributions?

Step Out

We know our own sources of power and joy better than anyone, *and* additional resources can shed further light. Personality assessments can be useful for naming our strengths—as well as the things that are decidedly not. So can people who know us well. Here, you'll draw on both. (*20–30 minutes*)

+ Begin by dusting off old assessments you've taken. Better still, take a new one or two. I'd recommend the RHETI test from the Enneagram Institute, Asia Suler's Earth Healing Archetypes, and the Green New Careers quiz—all linked at climatewayfinding.earth.
+ Then, reach out to 3–5 people who know you well (e.g., friends, mentors, coaches, coworkers, teachers). Send a short text or email to ask each of them: *As someone who's observed me in action, what do you think my strengths are? Where do you see me light up or shine?*
+ When you hear back, explore those responses alongside your assessment profiles. Jot down some notes: key words and phrases, what feels true or illuminating, and any strengths that stand out.

If you discover anything particularly compelling, you might want to add it to your map of power and joy.

Gather

This is a 75-90 minute experience.

Reminder: Everyone should bring their power and joy maps.

Tunes (*~10 minutes*)

Play songs from the playlist as your group gathers.

Poem (*~5 minutes*)

Read "An Old Story" aloud. Then, invite a second reading.

Opening Circle (*10–15 minutes*)

Begin an opening circle with this prompt—modeling first, then popcorning.

> In the spirit of honoring each other's superpowers, share one thing you've learned or appreciated from someone else in our time together so far. (And we'll also honor that we could each share many more than one!)

Reflection (*~5 minutes*)

Ask the group to read through their freewriting and maps from this chapter—and jot down any additional thoughts.

> What did you uncover about authentic power? What did you uncover about deep joy? If they overlap, is there significance in where the two meet?
>
> What stands out on your map? What's here that you want to offer our world or center in your climate contributions?

Discussion in Trios (*~15 minutes*)

Invite everyone to form trios and share. As ever, keep time and let people know when to switch.

Discussion as a Group *(15–25 minutes)*

Return to a whole group, and invite people to share from their trios. If useful, bring in these prompts.

> In your exploration of power and joy, did anything surprise you? Or feel challenging for you?
>
> What was especially illuminating for you in this chapter?

Closing Circle *(5–10 minutes)*

End with a closing circle using this prompt. Invite someone to start; then popcorn.

> Share one superpower you want to boldly claim for your climate engagement.

Closing Quote *(1–2 minutes)*

Read an inspiring passage from this chapter.

6

Connecting with Community

Rise to the lookout once more.
Seek your people. Sense your scene.

How did we humans find ourselves in the midst of a climate crisis?

The story goes something like this . . . Since the Industrial Revolution's debut, ancient subterranean carbon has been disinterred and put to use—a factory-kindler, a wheel-turner, a heat-maker, an electrifier. The burning of fossil fuels has been exponential, and the destruction of ecosystems has followed the same track. With every acre of forest razed, the carbon it once held gets released. All of it powers colonial expansion and capitalism, forging fortunes for a few along the way.

Forfeited to the atmosphere as pollution, that vast movement of carbon has now brought the world to the brink. The task at hand? Keep carbon where it belongs—underground and in living things—and bring some of that misplaced matter back down to Earth. We need to return carbon's *net* to *nought*, before it's too late.

This tightly abridged version of the climate story leaves all sorts of details out, but it illuminates the way our narratives often cast carbon as a villain or, at least, a real problem child. We hear about a *decarbonized* or *post-carbon* economy as the big goal. In other words, the paragon is carbon *gone*. And, of course: We must phase out the use of coal, oil, and gas. Period, full stop. The physics of our planet demands it.

But these typical portrayals of carbon can also limit our view. Agency does not lie with inert material, tucked below ground for millions of years. It sits squarely with the companies and governments extracting and building economies around it. More than that, though, if we only see carbon as something bad—a substance to be counted, capped, curbed, contained—we miss an opportunity for deeper, wider understanding.

There are other sides to what's called the *element of life*, and called that for good reason.

All carbon on Earth is the offspring of star burn and star scatter, vintage remnants refashioned on our blue-green planet. While carbon can certainly shine on its own—as pencil-pointing graphite or proposal-glinting diamond—it isn't actually all that keen on a solo act. It forms more compounds than all the other elements on the periodic table combined.

Made to commingle and bond with others, I imagine carbon atoms lonely and looking to turn *me* into *we*. Carbon soars airborne with oxygen until the fated photosynthesis breakup. Then, it hitches to hydrogen, also newly available, to stitch the greenery of lily pads and palm fronds.

That good witch of wood. Feeder of beasts. Infuser of soil. Ceramicist of shells. Whereas poet Walt Whitman wrote "I contain multitudes," in this case, it's the multitudes that contain carbon. Carbon gets together, teams up, and makes more than is possible on its own.

What if carbon—the element that connects and collaborates with such seeming ease and in such abundance—is as much something to learn from and emulate on our own climate journeys, as it is something to reduce, remove, or cold-shoulder? For carbon, relationships are the heart of the matter, the matter being all of life on Earth. A simple but powerful insight of social-impact strategists Katherine Milligan, Juanita Zerda, and John Kania rings in my ears: "Everything we know about systems tells us relationships are the core."

Much like carbon, we do not operate solo. That's true from our moment of conception, and reaching back for generations of ancestors before that. It's true inside—sharing our body with trillions of microbes. It's also true outside—relying, as we do, on everything from the outbreath of flora for oxygen to the food system's many hands for nourishment. One definition of *life*, according to writer and biologist Dr. David George Haskell, is *interconnection*. Life in a solitary state isn't life at all.

> Relationships are connections, and connections weave separate parts into a whole. Take stock: In what ways do you feel looped into something bigger?

As people, we are all enmeshed in our unique ecosystems, both literal and metaphorical. That's why, in *Climate Wayfinding*, we adopt a "carbonic" perspective and intentionally survey the contexts and communities that we are nested within and are a part of—or could be. We look to the many spheres, settings, affiliations, and associations that inform and enable our participation in planetary healing. We suss the relational webs where our paths cross with others' and collaboration takes hold.

Another writer-biologist, Janine Benyus, coined the term *biomimicry* to describe "conscious emulation of nature's genius." She and others in the field have identified core principles for mimicking the biological strategies found in wider nature, to solve the challenges we encounter as humans. One is this: "Be locally attuned and responsive." Life always solves for its needs *in context*, using materials and energy at hand, leveraging cyclical processes, and cultivating cooperative relationships.

If we were to engage in *Climate Wayfinding* without attention to context, community, and the threads between us, that wouldn't just be unwise or ineffective. It would be out of step with how life itself finds a way forward. And that is the very thing we want to embrace and embody.

Our Place in Context and Community

When facing a planetary crisis, as with a personal one, it's best not to go it alone.

Finding those we want to link arms with is the first step of building the biggest, strongest *we* possible. We need collective strength to counter formidable interests that wish to maintain the planet-unraveling status quo—typically those benefiting the most from it financially. We also need it because too many people are standing by, fingers crossed, waiting to see how things play out.

When asked for his take on *What can I do,* longtime journalist and activist Bill McKibben always offers some version of the same guidance: "The most important thing an individual can do is be less of an individual. Join together with others in movements big enough to matter."*

To put it another way: Be like carbon. Offer your atoms to a partnership, project, alliance, or campaign. Tap into the power of being more than one. It's possible an invitation is already in hand.

This is a strategy for results *and* for resolve. Again and again I have witnessed the power of community to get us to better ideas through a greater intelligence, to celebrate positive steps forward, and to commiserate and keep us going when the work is hard. Even when we don't bring our desired outcomes to fruition, the "mycelial" network can grow stronger, preparing for emergent possibilities. We need each other for all of it.

> When have you encountered climate connection or community? What made it so?

In December 2015, I collided with one of the most difficult moments in my adult life. Two leading lights in my family and local community had passed that autumn. Those losses converged with a painful professional impasse that left me feeling wholly unmoored from my sense of direction, agency, and passion.

*Bill McKibben was our very first guest on *A Matter of Degrees* back in 2020. Find that episode, "Give Up Your Climate Guilt," at climatewayfinding.earth.

Many months before, I had applied to attend a retreat for "young leaders and activists" hosted by the Center for Courage & Renewal. It somehow landed on the calendar at just the right moment. Emotionally flatlining and careening simultaneously, I headed exactly due south of Atlanta, to a modest lakeside retreat center.

Of the group of mostly twentysomethings, I was far from being the only person feeling lost or bone-weary. Thanks to the artistry and warmth of the facilitators, including teacher-author Dr. Parker Palmer, our weekend offered a kind of refuge.* The flow of introspection, reflection, and sharing helped to reground us in the truth of ourselves. It breathed new life into our participation in many different dimensions of social change.

Though my body was not in poor health when I arrived, my spirit was. I look back on that experience as one that, in an inner sense, saved my life. As the author and social critic Dr. bell hooks writes in *All About Love*: "Rarely, if ever, are any of us healed in isolation. Healing is an act of communion."

During the retreat, I learned about Parker's concept of a *community of congruence*, comprised of like-minded, like-hearted people—those who feel aligned and kindred, beyond blood relations. The group that convened for those three days felt like that kind of community. We came from many different places, professions, and lived experiences, yet there was a mutual encounter with being seen, understood, and inspired by one another. To be sure, we didn't all see things the same way, but we were dancing in the same direction and keen to dance, however briefly, together.

I left with a resolve to cultivate a community of congruence in my life and work much more intentionally. Certainly, that would include those I already knew well and felt connected with—friends, coworkers, teachers, partners in good

*I was very moved the first time I read Dr. Parker Palmer's small, wise book, *Let Your Life Speak*. It helped me see that our lives are calling us toward truth and wholeness in who we are and how we spend our days—and sometimes that message comes in the form of burnout or dead ends. It stirred courage to be my full, undivided self in the world, claiming my authentic place in the assembly of life.

trouble, and nonhuman soulmates too.* One of the first things I did back at my apartment was to map out that community.

I realized, as well, that following an inkling of shared values and visions might lead me to others I didn't yet know at all. It has and continues to do so, fundamentally orienting my own climate engagement, as I have moved toward those I most want to learn from and create with. The energy of relationality is what turned a book idea I'd conceived into a thing with pages: the *All We Can Save* anthology that Dr. Ayana Elizabeth Johnson and I co-edited in 2020, weaving the contributions of some sixty women. As she's advised: "Go . . . where your heart can find a home."

In the weeks following that retreat, I came to see that letting relationship lead can even transmute an impasse into a portal. From the anguish of a dead end, I redirected toward kindred community, and from that redirection grew new work and partnership. It fundamentally recentered and re-spirited my climate journey.

> Imagine navigating—finding your next step—by the warm pull of relationship. Who might you turn toward? Where might that take you?

Closely related to being knit into communities—those we join, those we build, and those we inherit—we exist within many rings and layers of context. Think of these as the surroundings and systems within which we operate, at various times or in various ways, ranging from the micro to the macro. Our contexts can constrain our choices, but they also can open up opportunities. Looking outward and understanding them can help us identify potential footholds for climate action and participation.

* The phrase "get in good trouble, necessary trouble" was a favorite of Congressman John Lewis, civil rights leader and long-serving public official, whose Atlanta district I have lived in for most of my life.

Geography is an immediately apparent context: our neighborhood, city or town, state or region, country, continent, hemisphere. Each of these brings with it physical, cultural, and political contexts too. An ecological view might highlight our watershed, our bioregion, our growing zone for gardening or farming. For those indigenous to or especially connected with particular land, that may be the most integral context of all.

There are also institutional contexts—for example, our campus, workplace, or place of worship. There are myriad professional sectors and spaces, social networks and groups, and civic associations and coalitions to which we might belong or have ties. And there are the sweeping contexts of movements, capitalism, mass media, democracy.

My own contexts range from the watershed of the Chattahoochee River to the world of publishing; loose ties to my former habitat of academia and strong ties to a network of women in climate; a local pub where I am a relative regular and a system of electoral politics that I engage with in bursts. And I feel the especially weighty context of the United States. This is the country that has produced the most climate pollution over time and bears the greatest historical responsibility for our global trouble—a context that spurs me on and has shaped my climate engagement in key ways.

Not every relevant context will be a space where we want to "do" climate. Besides, it would be impossible to be meaningfully active in all of them. But we can use the perspective of context to sense for the scale of action that might be right for us, from local efforts to international initiatives. We can discern whether we want to play inside the systems of government or business, put pressure on them from the outside, or eschew them, as best we can, in favor of forging something new. We can work out whether there are immediate needs or opportunities in our community calling for our response.

Our contexts also inform who we are, how we see the world, and how we experience and engage with it. The particular roles and positions we each have in society and in relationship to privilege and power—the hierarchical, structural kind—these can shape our climate engagement significantly. If we consider the dynamics of our citizenship, race, gender, sexuality, language, education level, access to wealth, bodily ability, neurodivergence or -typicality, and more, we awaken to biases we might bring with us. We also awaken to solidarity we might rally and resources we might marshal. As civil rights scholar Kimberlé W. Crenshaw reminds us, the intersections of all these things are fundamental and potent.* To be sure, none of that is the totality of who we are, *and* it is all inescapably here with us.

I imagine each of us standing at the convergence of our unique surrounding contexts, playing out on and within our very beings—like ripples from many directions meeting in the same body of water. Bringing awareness to these dimensions of self-in-context can help us be more thoughtful and choiceful, which feels especially important in climate relationship and climate community.

In my experience, it is actually the multiplicity of our contexts that makes the *we* strong. By bringing our diverse lenses together with others', we begin to create a much more powerful kaleidoscope of perception and insight than we could ever achieve alone. Where we might be tempted to draw lines of difference, we have an opening to move humbly together instead, though we may bumble or blunder along the way. I think of lichen—that unlikely partnership of fungus and alga. By merging their gifts, they bedeck the planet in silvery green, sunburst orange, the black of a moonless night.

> Gold found within multiplicity. Footholds found within context. Notice how these notions land or stir in your body.

Is it possible to invite our differences to coexist, while also finding the congruence we crave? In my experience, yes, if shared *values* create a current of coherence. But that requires clarity. Values often have a bent for moving namelessly in our shared waters, yet coming to the surface in pivotal ways.

*Kimberlé W. Crenshaw coined the term *intersectionality* in 1989 to express how various aspects of one's identity can combine and compound, with more complexity than 1 + 1 = 2—a phenomenon she illustrated through the experiences of Black women.

Values are fundamental beliefs about how we live our lives, or should. (See "Values for Climate Healing.") At their best, values become principles that guide us day-to-day and through our hardest decisions, both individually and together. At their worst, the values we proclaim are brittle under pressure, and there is discontinuity between word and deed. I have few regrets, but those I do have are the traces of such breakage.

The values we hold, whether implicit or explicit, often flow from our contexts and communities—our family, culture, education, faith tradition, and more. They imbue us with foundational beliefs, essential ethics, cosmologies, and ways of understanding or being with reality. While we are heirs to those, we also have the ability to update them, maybe releasing some, maybe adopting others. As tribal attorney and land defender Tara Houska (Zhaabowekwe) urges, we need to embrace "values that reflect life."

Because a core truth of the living Earth is profound interdependence, values like kinship, mutuality, and reciprocity are both our birthright and birth-responsibility as human beings. We need them not only to animate how we move through our lives but also to counter the widespread values that are wildly out of alignment with life—such as greed, superiority, me-and-mine above all else. Māori writer and activist Nadine Anne Hura reminds us that these are the fundamental causes of the sickness we are experiencing on Earth. They are the values that have shaped the global economy as we know it and spawned all that unearthed, misplaced carbon.

Values, then, are a bridge between the inner and the outer. We hold them internally, but also with others. They shape how we move through the world around us, and the world around us shapes the values that we hold. By clarifying our guiding principles and amplifying them in action, we may find ourselves better able to commingle and flock, like migrating birds who travel as individuals yet together—collective intelligence embodied on the wing.

"There is an art to flocking," activist and author adrienne maree brown observes, "staying separate enough not to crowd each other, aligned enough to maintain a shared direction, and cohesive enough to always move toward each other." We flock through reciprocal, responsive awareness of ourselves, our partners in flight, and a destination calling, *Yes, beauties, this way.*

> What values do you hold and embody that can help us flock well—or turn *me* into *we,* as carbon does?

The story of carbon is a layered thing, as we have seen. And that layered story is still being written—as we gain views into the past, make sense of the present, and, from there, begin edging further on and shaping new arcs that honor life.

Backward is where the scientific technique of *carbon dating* takes us. How old is a piece of bone or wood or linen? Up to about sixty thousand years, carbon can tell us. Over the course of a lifetime, plants absorb and animals ingest carbon. A smattering of that carbon is radioactive. Unstable in form, it shapeshifts over centuries and millennia, decaying at a predictable rate. Carbon dating is essentially a fancy counting exercise of how much of that *radiocarbon* remains. If you know that, you know age.

This technique, first developed in the 1940s, has helped us to understand the full carbon cycle—how carbon moves between sky, sea, soil, living organisms, their buried remains, and back again. It has also helped to affirm the anthropogenic origins of climate change. Fossil fuels are *fossils,* many millions of years old, so their fraction of radioactive matter has long decayed away. That absence makes it possible to identify their gaseous remains in the atmosphere—like a fingerprint for carbon that could only have come from prehistoric gas, oil, or coal, burned by human beings. Carbon thus reveals herself.

If carbon dating is about putting a fine point on the past or the conditions of the present, I am taken by the idea that *carbon dreaming* may be its forward-looking correlate.

To dream is to move into a realm of the unknown. At night while we sleep, we journey through images of this, sensations of that, storylines that jump from here to there and back again. And dreams themselves, their source and purpose, remain quite pooled in mystery.

Perhaps no realm is more unknown than the future—every inch of it pending revelation. In the context of the climate crisis, that unknowableness, unpindownableness, is often a source of angst. *What might befall this place, these people? Will we, and our loves, survive?* But as much as it may beckon us to nightmare, the unknown that lies ahead may also spark our hotfoot imagination, even untamed fantasy, of longed-for possibilities coming to be.

The in-process story of carbon is its own, yet it is also scribed by every other life—an autobiography held in common and unfolding in real time, in our skies and in ourselves. Carbon is ultimate context and elemental community and emblematic mutuality. And it is a reminder: *How will we survive, transform, flourish?* Only together.

If the future lives anywhere, it must be between us. In my experience, it can live between as few as two people, who conjure a quippy podcast, plan a pollinator garden, or conceive that an old rail line could change the way a city moves. Better still if the future lives among throngs of us—growing each time we dare to have that vulnerable conversation, each time we step toward that daydream that's begging and bigger than ourselves.

For the uncharted expanse ahead, my own carbon dream is this: We will be prisms, joined one to the next. And we, in our relatedness, will bend the line between shadow and light.

Values for Climate Healing

To get wheels turning, this is a (non-exhaustive!) list of values that may signify what matters most to us and guide how we engage with the Earth, others, and ourselves.

Abundance + Accountability + Active Hope + Ancestry + Attention + Authenticity + Balance + Beauty + Belonging + Biophilia + Boldness + Care + Clarity + Collaboration + Community + Compassion + Consciousness + Courage + Creativity + Curiosity + Deep Listening + Determination + Discovery + Dignity + Diversity + Dreaming + Embodiment + Emergence + Empathy + Enough + Equality + Equity + Evolution + Exploration + Fairness + Forgiveness + Freedom + Future-orientation + Generosity + Global-mindedness + Grace + Gratitude + Harmony + Healing + Heartfulness + Honesty + Humility + Humor + Imagination + Improvisation + Indigeneity + Inspiration + Integrity + Intentionality + Interconnection + Interdependence + Intuition + Joy + Justice + Kinship + Legacy + Liberation + Localization + Love + Loving-kindness + Loyalty + Mutuality + Nonviolence + Nurturing + Oneness + Patience + Peace + Perseverance + Play + Pleasure + Possibility + Power-with + Presence + Purpose + Reciprocity + Redistribution + Relationality + Remembering + Renewal + Repair + Resilience + Resistance + Responsibility + Rest + Reverence + Rigor + Rootedness + Safety + Sanctuary + Showing Up + Simplicity + Slowing Down + Solidarity + Sovereignty + Spirituality + Stewardship + Transformation + Transparency + Trust + Truth + Understanding + Vitality + Well-being + Wholeness + Wildness + Wisdom + Wonder +

the mississippi river empties into the gulf

Lucille Clifton

and the gulf enters the sea and so forth,
none of them emptying anything,
all of them carrying yesterday
forever on their white tipped backs,
all of them dragging forward tomorrow.
it is the great circulation
of the earth's body, like the blood
of the gods, this river in which the past
is always flowing. every water
is the same water coming round.
everyday someone is standing on the edge
of this river staring into time,
whispering mistakenly:
only here. only now.

✲ Lighting the Way

with Wanjira Mathai (Nairobi, Kenya)

The kindred community we conceive, cultivate, and journey with often constellates around our contemporaries. But it can also reach beyond the boundaries of *now* into ancestry and lineage—and may be at its best when it does. The power of holding common purpose across generations is at the heart of Wanjira Mathai's path.*

Wanjira grew up in Nairobi, Kenya's bustling capital city, and spent formative holiday time on her grandparents' farmland. When she was six years old, her mother, Dr. Wangari Maathai, started a groundbreaking grassroots movement to mend the impacts of deforestation across the country.

"Nature is the source of everything good," my mother would tell me all the time. In our neighborhood, there wasn't much space to plant trees, but we did. We planted them everywhere. We surrounded ourselves with green—always you had to plant something. People could find our house by following the trees.

After high school, Wanjira wanted to go her own way. Her studies in biology and then public health brought her to the United States, and she joined the Carter Center's efforts to eradicate rare and neglected diseases around the world. It was deeply meaningful work, but a few years in, she began to sense a crossroads.

I felt a motivation to do something different, but I didn't know exactly what that should be. Eventually, I decided to go home and take a break. My intention was to return to the US after a year. But while I was back in Nairobi, my mother asked me to help with the movement she had been building.

*In 2021, Dr. Ayana Elizabeth Johnson and I interviewed Wanjira Mathai for a joint episode of our two podcasts. Find that episode, "How Gender Equality Can Save the Planet," at climatewayfinding.earth. ⊘

In 1977, Dr. Maathai founded the Green Belt Movement in response to women across Kenya who had shared with her the impacts that deforestation was having on their lives. As the trees went, so did firewood, food, shelter, water, and the ecosystem stability that trees provide. The aim was to turn around those entangled trends. Attending a meeting here, helping on a proposal there . . . before she knew it, Wanjira was steeped in the mission with her mother.

As I got involved in the work, I began to see, wow, this is an amazing thing that's going on here. A network of thousands of women, organized into groups, was growing and planting millions of trees and being compensated to do so. The Green Belt Movement is about communities mobilizing to restore landscapes, but also to restore dignity, secure livelihoods, and claim their rights. Planting trees is the entry point.

Wanjira was still planning her return to the US when, in October 2004, Dr. Maathai became the first African woman and first environmentalist to be awarded the Nobel Peace Prize.

Every so often I go back and read the Nobel Committee's citation. It was not only an award for the Green Belt Movement. It was an award to her for having made the connection between the environment, democracy, and peace. There was such clarity in her understanding—how intricately linked they are.

"We are called to assist the Earth to heal her wounds and in the process heal our own," Dr. Maathai said in her Nobel lecture, "indeed to embrace the whole of creation in all its diversity, beauty, and wonder." But the connection between the environment and peace was very progressive at that time, and far from obvious for the broader public. Explaining it, with the Green Belt Movement as her deep-rooted example, became Dr. Maathai's central work until she died in 2011.

Wanjira never left Nairobi and never left the work. She now sees something fated in her return—a gift to have the chance to collaborate with her mother for a decade. In the process, that way of seeing the world was passed viscerally from one generation to the next.

When someone says "sustainability," I immediately see a web. And whether you call it sustainability or regeneration, what really matters is the connectedness of different issues and aspirations—climate, democracy, sustainable development. All of these linkages are essential.

Wanjira now brings that perspective to her leadership of the World Resources Institute's work in Africa.

Of the fifty most climate-vulnerable countries in the world, two-thirds are found on the African continent, which is also a hub of solutions. Vulnerability and opportunity—both are present.

At the World Resources Institute, we are part of an Africa-wide initiative to restore more than 100 million hectares of land, which is at once about curbing climate change and *cushioning communities against its worst impacts. I envision flowing waterfalls and rivers, thriving bird life, a continent that is able to feed itself, people living dignified lives.*

This vision we have to re-green the African continent, it's not only possible; it's vital to our life support system.

Core to realizing that possibility, Wanjira reminds us, is trusting and supporting communities. Community is the essential context, catalyst, and beneficiary of climate healing. In what may seem to be a contradiction, small is the only way to reach such a massive scale.

To turn the tide requires that we work in partnership, that we catalyze others, that we invest in those who know best. In the end, all solutions are local because it all happens on the ground. Our research shows that when efforts are locally led and managed, they are twenty times more likely to deliver long-term success—both environmentally and economically.

But global solidarity is also essential. Our global priorities must grow from the realities on the ground, and investment must follow. There must be a constant conversation up and down.

Building that bridge from the grassroots to a global community. Cultivating restoration champions across Africa. Creating access to clean, affordable, and reliable electricity. Unlocking prosperity, especially for women and youth. Across her wide-ranging and interconnected work, Wanjira continues to walk in what she describes as her mother's light.

My mother was working at the exclamation point of her passion. And she often spoke about persistence, patience, and commitment. This is extremely slow and deep work. We shouldn't be overwhelmed by the enormity of it. If I am doing my best—with what I have, from where I am—that is enough.

The foundation we started in her name works with young people to build a culture of purpose and integrity. Courageous leadership is what we need today. Fostering it—that's now my passion.

When I look to Wanjira, I can't help but think of monarch butterflies on my own continent and their knowing northward movement to milkweed—a 2,500-mile migration, just one way. The route is so long that it can exceed a single monarch's lifespan. It will be great-grandchildren or great-great-grandchildren that complete the full cycle of the journey—by some passed-down secret, knowing how and where to go. We, too, carry our ancestors' dreams.*

*Wanjira's story is rewoven from an exchange we had, as well as other material. Hear more from her at climatewayfinding.earth.

Playlist

Songs to grow nested and connected.

"IKYK"—Ogi

"Crowded Table"—The Highwomen

"Call It Dreaming"—Iron & Wine

Tune in at climatewayfinding.earth.

Map

Engage again with creative cartography to explore the many contexts you exist within and to reflect on your own kindred community. The prompts below will guide you to create two maps. (*10–15 minutes*)

What you'll need: two two-page spreads in your journal or two large sheets of paper; a pen, markers, and/or colored pencils.

Three notes before you start:

+ Don't filter. Include anything or anyone that comes to mind, even if they don't seem connected to climate. (If you need additional clarity on the concepts of *context* or *kindred community*, zip back to the essay.)
+ Playful creativity encouraged—let your inner artist have some fun.
+ If you already feel very connected to community, this activity may feel sweet and enlivening. But it can also feel challenging and even achy. If that's true for you, please know you are not alone. Many of us are seeking more and deeper connection, and *Climate Wayfinding* can be a step toward that.

If you'd like to see example maps for inspiration, find some at climatewayfinding .earth.

These prompts will take you, step-by-step, through populating your maps of context and community. As something pops into your mind, jot it down. (You might want to play music while you map.)

Put your name in the center of each map.

Begin with mapping your *contexts*—the surroundings in which you operate at various times or in various ways, from the micro to the macro.

Consider:

What are your geographies and places?
Your groups and networks?
Your educational or professional sectors and spaces?
The larger systems around you—ecology, culture, politics, movements, economy, etc.?

When your context map feels complete, go to your second spread or sheet of paper and begin to map out who is part of your *kindred community.*

Consider:

Who are like-minded, like-hearted people in your life?
Other creatures or wild beings?
Who feels aligned and congruent?
Makes you feel seen, understood, and inspired?
Seems to be dancing in the same direction?

You may know these folks very well or barely at all, but you have a sense of shared values and connection with them.

When you feel complete, take a minute to look over both maps. Circle or star what and who feel most relevant to your climate engagement. Then, shift to the journal prompts below. (If you're participating in a group, bring your maps to the next gathering.)

Journal

Take a moment to look back over your maps of context and community. Then, begin to freewrite on the first prompt.

> What insights surface as you reflect on yourself both in context and in community? If you were to center relationships, what possibilities might open up for your climate engagement?

Now, read the prompt below, and take a minute to close your eyes, breathe, and see what comes. Then, put pen back to paper.

> Reflect on the values you've inherited and those you've cultivated over time. Write down 3-5 core values you hold. What do they each mean to you? How do you express them?

Step Out

Community is a garden. These steps will help you cultivate yours. (*20–30 minutes*)

+ First, reach out to at least one person from your map of kindred community. If you're already in the depths of climate collaboration, you might focus this on *appreciation*—thanking them for the role they've played. If it's someone you haven't "done" climate with yet, you might focus this on *exploration*—inviting them to have a conversation about what good trouble you might get into together. Approach this in whatever way feels authentic to you.
+ Second, scout for climate projects, campaigns, or groups that feel aligned. Move with the speed of a brainstorm and generate a list as you go. Once you have a solid list in hand, choose one project/campaign/group you want to connect with. Sign up for their email list or an upcoming event or engagement opportunity. You might also reach out to team up, offer support, or help amplify their work within your own networks.

If you're not sure where to begin scouting, there are some resources for you at climatewayfinding.earth.

Gather

This is a 75-90 minute experience.

Reminder: Everyone should bring their context and community maps.

Tunes (*~10 minutes*)

Play songs from the playlist as your group gathers.

Poem (*~5 minutes*)

Read "the mississippi river empties into the gulf" aloud. Then, invite a second reading.

Opening Circle (*10–15 minutes*)

Begin an opening circle with this prompt—modeling first, then popcorning. (You might want to write these offerings down so people can come back to them.)

> Communities thrive because of the many gifts people bring to them. Share one thing you would love to offer this community.
>
> It could be a characteristic you possess, knowledge you hold, or a skill or resource you can offer. Anything you feel moved to give.

Reflection (*~5 minutes*)

Ask the group to read through their freewriting and maps from this chapter—and jot down any additional thoughts.

> What insights surface as you reflect on yourself both in context and in community? If you were to center relationships, what possibilities might open up for your climate engagement?
>
> Reflect on the values you've inherited and those you've cultivated over time. Write down 3-5 core values you hold. What do they each mean to you? How do you express them?

Discussion in Trios (*~15 minutes*)

Invite everyone to form trios and share. As ever, keep time and let people know when to switch.

Discussion as a Group (*15–25 minutes*)

Return to a whole group, and invite people to share from their trios. If useful, bring in these prompts.

> Where have you seen community (climate or otherwise) fostered well, across lines of difference? What helped make the *we* strong?
>
> Were there aspects of this chapter that made you feel seen? Or made you feel stretched?

Closing Circle (*5–10 minutes*)

End with a closing circle using this prompt. Invite someone to start; then popcorn.

> Share one core value you hold dear and bring to your climate engagement.

Closing Quote (*1–2 minutes*)

Read an inspiring passage from this chapter.

7

Orienting with Vision

Turn your gaze forward now
with direction from the heart.

"The task of stories is to teach us to see," Dr. Margaret W. "Jane" Pepperdene said from her lectern. "To see with a more perceptive inner eye. To see feelingly."

It was my second morning in Jane's classroom, where, as a professor emerita of English, she taught just one high school seminar course each year. (Famously, she didn't change a thing about her teaching or expectations of students, who would simply need to step up and meet her.) We would spend the next year studying "The Figure of the Journey," from Chaucer to Eudora Welty, Shakespeare to T. S. Eliot. We would read stories and poems of pilgrimage, that purposeful movement toward a particular place of moral and spiritual value. I would come to see the way we are all journeying toward meaning in our lives.

I remember Jane's warm, oaky timbre—sometimes soaring, sometimes a near whisper—and her spirited hands—moving, pausing with intent, and moving again as she lectured. We followed her down and down into the heartwood of a text. Her questions called us into dialogue with it. I can still sense the tears that would spill onto my cheeks on occasion, so moved by how stories create openings to find ourselves and one another, how they beckon us toward something fuller, more genuine.

It turns out that the ordering and sensemaking we do through story resides in the same section of the brain as our capacities for navigation. The hippocampus, so named for its seahorse shape, is hardworking gray matter that crafts autobiography and makes mental maps. (It's famously large for long-tenured London taxi drivers, who have maneuvered the city's maze for decades—pre-GPS.) The hippocampus also plays a crucial role in the process of imagination, as we project into the future and envision our goals and aspirations.

The capacity to picture a place we wish to go—to conjure *there's something good over yonder, onward*—is core to navigation. In her book *Wayfinding*, journalist M. R. O'Connor writes, "The human imagination is like a beacon that orients us, helping us to make decisions about where we want to go and how we might get there, as well as self-regulating our behaviors and emotions in service of a destination or a destiny." Imagination is something of a lighthouse, illuminating from within.

Yet that lighthouse can feel hard to access amid the churn of the climate crisis and all the rest. We may find ourselves worn flat out, just trying to wrap our mind-hearts around the current contours we face. To invoke another reality? To imagine our world recast and radiant? To see our role in forging it *and* a way through to get there? That can feel daunting, undoable, even absurd. If we want movement forward, though, we must first turn our gaze there. We must invite vision into the story.

> Is there a place—past, present, or pictured in your mind—that you can "see feelingly"?

Vision, in its most common usage, refers to the ability to see—that collaboration of the eyes and brain, receiving patterns of light that bounce off objects and then turning those patterns into images. Reflected photons make known to us the dance of aspen trees, wild irises' pops of purple-blue, or an evening herd of elk. In *Climate Wayfinding*, we key into another dimension of vision, too: the capacity to look to the future with imaginative wisdom.

In this sense, vision is about where we want to go—catching a glimpse of our climate engagement to come, the scenes in which we might participate, the world we could manifest together. This is about what lies ahead, sketched in mental images we hold in the present moment. Those images can be a summons to ourselves, enticing us to take steps in a particular direction.

To cultivate this kind of forward-looking vision, we rely on the light within, rather than the light of an LED bulb or the sun. It's still about seeing, but not with the physical eye.

As a coach and advisor, Nikki Silvestri teaches that vision sources from our spirit or soul, flows into the feeling realm of the heart, moves into the mind, and then expresses into the physical world. I have been a beneficiary of that insight. For me, it translates as vision moving something like water: It burbles and springs from our core, then percolates to the surface of our being, and finally ripples or rushes out into the world beyond. From barely visible headwaters, a river arrives.

In the fall of 2021, at Nikki's urging, I undertook a visioning exercise—first going inward in meditation, then letting that pour onto the page to document what I saw for my future self. I stepped into the life I might be living and the work I might be doing, three years further on.

This is some of what I wrote: *I am working to shift the story of humanity on Earth, using various means of voice, and to nurture participation in that shared story through learning exchanges and community building. At The All We Can Save Project, we have launched a course/program to help people find and deepen their place in climate—to gain clarity, transmute grief into fuel, have a sense of what's possible and a sense of belonging in the birthing of that possibility. For my part, I am living and leading more and more like a tree—with greater groundedness, generosity, and grace. I am going deep so I can grow wide.*

Climate Wayfinding didn't yet have a name. Three seasons would pass before running a pilot of the program. But it was there, already springing or already sending down a taproot.

Beyond the specifics of what I glimpsed for myself at that time, I registered something essential about our human capacity for perception. We don't need to be prophetic mystics or ace entrepreneurs to source original and guiding ideas, and vision needn't be about mountaintops or masterpieces or quantum leaps. The brain, in all its ordinary wizardry, can patch together elements that are familiar today into wholly new collages of tomorrow—assemblages yet unseen but beginning to crystallize in the mind.

The imagination wants room to roam. If we absolve any expectations to see clear contours of what's ahead—if we are open to receiving whatever comes—we can engage with vision not as a weighty, serious task but as creative play. *So what if we circle back and scan again?* It's possible that *re*imagining is actually essential practice. I now see our figments, reveries, and dreams as gifts to our current selves. When the mind ventures forward, using that more perceptive inner eye, it releases us from the constraints of today's reality, however momentarily.

As we begin to turn the bend toward what might come beyond the book in hand, we tap into this particular form of ingenuity and expand our ways of knowing about where we could go from here. Vision might just unveil a navigable channel or point us to a crossable mountain pass.

> Take the pulse of your imagination. Give some compassion to any apprehension, and give some encouragement to any readiness for creative play.

In my experience, as we move in the direction of vision, our path appears, disappears, and catches our ready soles once again. The vision itself may clarify, shift, or expand as we go. Given this undulating nature it can be useful to have some steadying tools at hand. Conventionally, a map has been one. A compass has been another.

The first compass was invented in ancient China, during the Han dynasty (206 BCE–220 CE). It looks something like a metal soup spoon on a square plate, which has lines running corner-to-corner and side-to-side and a circle in the center. In 2008, the opening ceremony of the Beijing Summer Olympics celebrated it as one of the Four Great Inventions, alongside gunpowder, paper-making, and printing.

This earliest compass, while directional, was not made for navigation. Its function was spiritual: to bobble, swivel, and point south, thereby guiding feng shui, the flow of energy within spaces and buildings. It may also have offered a means of divination. That is, the compass was a tool for creating more harmony and balance at home rather than trekking far away, and for bringing the self into contact with a future not yet visible.

Centuries later, during the Song dynasty (960–1279 CE), the compass was adapted for terrestrial and maritime navigation. From there it became a central tool for voyages to the far reaches of the world, for both good and ill. The compass enables us, as the word's Latin root suggests, to "step together." We can only speculate how different our world might be if a primary impetus for such stepping weren't imperial.

Ancient and modern compasses alike align to the poles of our planet, which is, among so many other things, a magnet—the ends of which sit some 7,900 miles apart if measured straight through the center. That magnetic field is created by molten iron and nickel roiling in Earth's core, and it extends as a vital protective shield into space. Rather critically, the magnetosphere shelters us from the bombardment of solar wind and storms. In other words, inner stirrings and inner heat enable this orb to be a living planet—and one a compass can gauge.

With a pointer ever in keen pursuit of magnetic north, the contemporary compass is key for those navigating backcountry trails or open waters. Even in dense fog or beneath overcast skies, a compass can make the path apparent. It can make the maps we have legible and arrival at our desired destination more likely.

As a child, I remember being fascinated by a small ball compass that hung from my grandmother's keys, always moving in its watery sphere. But I first learned to properly use a compass for navigation right around the time I journaled my Earth/life commitments at age sixteen. I recall laying out a well-worn topographic map on a flattish rock in Pisgah National Forest, placing the compass along the map's edge, and rotating until north on the map and the compass's north-seeking arrow found each other, flush.

Suddenly, the landscape around me revealed new information. I could discern that creek, that peak, that overgrown and unmarked trail. The topo map was largely useless without the compass, which was key to determine the right next steps to get back to camp and community. The compass helped me find my direction by bringing me into deeper relationship with the Earth.

In *Climate Wayfinding*, we adopt a compass that is, by design, beneficial, both individually and mutually. To support the ongoing cultivation of vision, and our efforts to track toward it, we use a framework called the Climate Compass. It helps to merge the many trails of our exploration—giving us a record of the terrain we have covered and an artifact to keep, but most importantly providing a tool to continue working with as we go. (See "Orienting to the Climate Compass.")

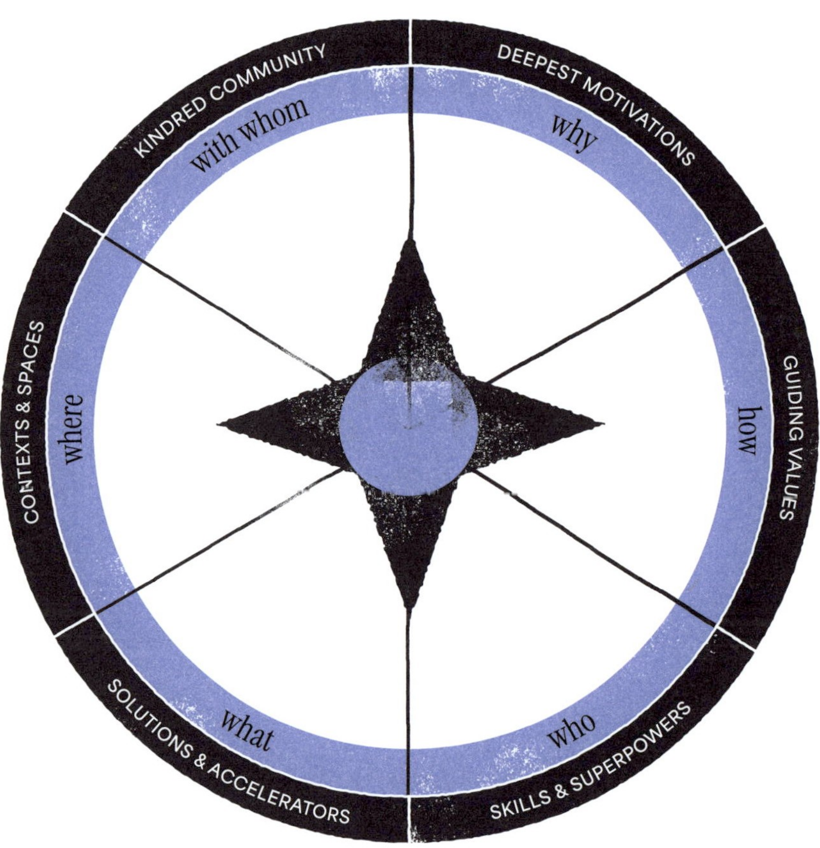

The Climate Compass

> Imagine an arrow within you, longing for north. What sensations does that arrow create in your body? Where does it point?

The idea of an *inner compass* seems to have arisen within multiple wisdom traditions. Psychologist and Buddhist teacher Tara Brach describes orienting from the "compass of the heart" as accessing the very deepest intentions we hold and then acting from them. It means, she says, awakening our full potential, wholeheartedly caring, and engaging in the here and now.

The Climate Compass, then, is a version of this fairly universal idea for our current moment on Earth. In a time where maps increasingly no longer work, we are reliant on the strength and refinement of our own inner compass, as well as connections with those around us who are well attuned and equipped.

Like the modern compass, this tool can assist us as we engage with our rapidly shifting planet. But it also reaches back to the purpose of its earliest kin. Like an ancient compass, it can help us make home, including the home of our own existence, more energetically true. And not unlike both of those devices, it guides by connecting to the Earth itself.

Of course, a compass holds little value sitting in a drawer or on a closet shelf. It begs to be in hand and on the excursion. So the Climate Compass holds an invitation: *Trust that there's an energy moving in and through you. Feel it and follow. Turn your feet, swivel your shoulders, and take one step forward, even if it's the only step that's clear.* (We will take up the specifics of those steps in the next chapter.)

I suspect a compass might, if it could, revolt against the idea of a final destination, as journey's end would also mean the end of its opportunity to play a part, to pitch in. And what would it do with all that longing to read the landscape? To orient and guide? Like any good circle, it's made to go around—to be turned, revisited, considered from a new perspective, in a new place. It's made to meet life in its actuality, which is never so simple as to land in one direction and follow it forever. Life is also curving and spiraling and cycling back around.

In this especially labyrinthine time, the path forward for our climate engagement will not be uniform or fixed. The human experience tends to be full of splits and forks, and that branching will almost certainly become more turbulent in this period of change. So we are all the more in need of a living instrument to help us get our bearings and guide our moves—something that might balance or brace us when the surrounding world buffets, even buckles.

The Climate Compass offers no assurance against getting lost, nor should it. Lostness is a fundamental experience of life. It can reconstitute us. It can lead us to places we did not know were options and, maybe to our surprise, wind up opening a fresh route. As philosopher-writer Dr. Bayo Akomolafe shares: "In order to find our way, or find ourselves in new ways, my [Yoruba] elders would say . . . you must become lost, *generously* lost. . . . This lostness is not a negative thing. It's how we stumble upon exquisite new worlds . . . new ways of being."

Chickasaw poet and storyteller Linda Hogan puts it this way: "Sometimes being lost, if there is such a thing, is the sweetest place to be."

Like any compass, we hold this tool alongside other, broader path-making influences. The very ground underfoot may croon, *This way* . . . Ancestors may whisper and shepherd. Happenstance may lift us into its peachy swirl. Maybe we will steer by the prayer in the wind's thirsty whip—or the whippoorwills' wilting trill—or the trilliums' untimely waking. Everywhere, there is emergent and earthly co-navigation.

> Do you sense other allies–ancestral, wild, numinous–that are with you on your path?

M. R. O'Connor's *Wayfinding* makes powerfully and beautifully clear that navigation is primordial for the likes of zebras, desert ants, and lobsters, but also, in unique ways, for human beings. When we navigate, we are leaning into something older and deeper—something spiritual as much as scientific.

We are constantly mapmaking in our mind's eye and constantly storying ourselves. *This is where we are. That is how we got here. Over there is where we are going. And here's why it all matters.*

I was living in England, completing the last slog of my doctoral thesis, when I got a call from a mutual friend that Jane, my beloved teacher, was very ill. We had stayed in touch through the years, often over tea in her living room. (She was delighted when I broke up with a college boyfriend, at least in part, because he thought literature wasn't "real." Incidentally, he made a career in oil and gas.)

I frantically wrote a letter and emailed it to my father for hand delivery, posthaste. I hoped to get back to Atlanta in time to tell her in person, but I wasn't taking any chances to express again my love, admiration, and gratitude for all that I had learned from her about the human journey and a life of meaning. *Beyond and yet inextricably rooted in the classroom, in so many ways I am who I am because you were and are my teacher.*

Jane exhaled her final breath that day, just before the letter could reach her. I moved home two weeks later, landing into the rawness of her absence—the absence of a mentor so central to how I'd found my way for a decade. Nothing can erase that kind of loss, but as weeks and months went by, I discovered how much she was still with me and, no doubt, the many students whose paths she shaped over a lifetime of teaching.

When I return, in memory, to Jane's classroom—where I was thoroughly afire—I discern more clearly that what we're doing in *Climate Wayfinding* is so fundamentally human. Our impetus for pilgrimage may be different than for Chaucer's bawdy band, telling their tales en route to Canterbury, or Eliot's ensemble of speakers, crossing the ruins of *The Waste Land*. In its specifics, it almost certainly is. But we are stepping into patterns that are ancient. That doesn't make them timeworn. It makes them trustworthy. There is something steadfast in the seeking and assurance in the act of excursion.

I find it simply remarkable that in the very makeup of our brains, we are built for the necessary work of our era. We have within us a seedbed of narrative, imagination, and navigation. We have within us a seedbed of the world we want to grow.

I step into an opening in the tree canopy to regard the night sky—there, piercing the inky dark, a distant starry seedbed. Below the boundlessness, I can see something, feelingly, for the first time. A compass is more than an instrument we hold. A compass is something we are.

Orienting to the Climate Compass

We've reached a moment of convergence on the journey, and the Climate Compass helps us organize insights gained along the way.

Each of the six segments anchors to essential content of prior chapters. These core areas of reflection and discovery are unique but united in the compass, a tool to accompany us on the path forward from here.

Why	Our deepest motivations for being engaged in climate healing, with roots in our emotions.
How	Our core values or guiding principles, which shape how we show up in climate engagement and beyond.
Who	Our key skills and superpowers, animating who we want to be in the ecosystem of transformation.
What	The climate solutions and accelerators for change that we form relationships with to remake the world.
Where	Our key contexts, spaces, and systems, where we gain footholds for climate action and participation.
With Whom	Our kindred community, those with whom we connect, collaborate, and make good trouble.

♥ Dead Stars

Ada Limón

Out here, there's a bowing even the trees are doing.
          Winter's icy hand at the back of all of us.
Black bark, slick yellow leaves, a kind of stillness that feels
so mute it's almost in another year.

I am a hearth of spiders these days: a nest of trying.

We point out the stars that make Orion as we take out
     the trash, the rolling containers a song of suburban thunder.

It's almost romantic as we adjust the waxy blue
     recycling bin until you say, *Man, we should really learn*
some new constellations.

And it's true. We keep forgetting about Antlia, Centaurus,
     Draco, Lacerta, Hydra, Lyra, Lynx.

But mostly we're forgetting we're dead stars too, my mouth is full
     of dust and I wish to reclaim the rising—

to lean in the spotlight of streetlight with you, toward
     what's larger within us, toward how we were born.

Look, we are not unspectacular things.
     We've come this far, survived this much. What

would happen if we decided to survive more? To love harder?

What if we stood up with our synapses and flesh and said, *No.*
     No, to the rising tides.

Stood for the many mute mouths of the sea, of the land?

What would happen if we used our bodies to bargain

for the safety of others, for earth,
if we declared a clean night, if we stopped being terrified,

if we launched our demands into the sky, made ourselves so big
people could point to us with the arrows they make in their minds,

rolling their trash bins out, after all of this is over?

✡ Lighting the Way

with Jennifer Robinson (South Coast of New South Wales, Australia)

Where should we point the compass of the world? In 2015, small island nations implored all countries to adopt 1.5 degrees Celsius as the boundary line for global warming. "1.5 to stay alive" was the harrowing slogan. Their efforts secured a long-shot victory: It was etched into the Paris Agreement, a legally binding treaty, which some described as extending a lifeline. But action over the last decade has been wholly inadequate, and now that goal is on the brink of being engulfed. For a group of law students in Vanuatu, the press of urgency produced a seemingly impossible vision: to buoy up *1.5 degrees* and give it legal brawn. Human rights lawyer Jennifer "Jen" Robinson found their vision impossible to resist.

When Jen was in graduate school, studying international law, she would have told you—as she told me at the time—that the work she intended to do in the world was decidedly distinct from climate. She was focused on human rights. Self-determination. Free speech. Fairness.

I remember when we would speak about climate, sitting in our flat or at a pub. I would say firmly: I'm a human rights lawyer, not *an environmental lawyer. But you challenged me: What if there's no planet for humans to live on? Our conversations helped me understand that the climate crisis* is *a human rights crisis.*

As Jen's practice got going, she decided to dedicate some of her time to climate justice. She took on a case protecting the right to protest fracking in the United Kingdom, and another that saw fracking policy declared unlawful on climate grounds.* In parallel, she was working with the government of Vanuatu, an archipelago to the east of her home country, Australia.

*Fracking, or hydraulic fracturing, is a way to extract gas or oil by injecting water, chemicals, and sand into shale rock formations. The high-pressure mixture cracks the rock, releasing fossil fuels. The harms are many, including water contamination and leaks of methane, a heat-trapping gas eighty times mightier than carbon dioxide.

There was an opportunity to see the Chagos Islands returned to Mauritius—part of broader efforts to support self-determination for colonized peoples around the world. The UK had controlled those atolls in the Indian Ocean since the 1960s. I worked with Ralph Regenvanu, then the foreign minister of Vanuatu, in the country's first appearance before the International Court of Justice (ICJ). The world court agreed with Vanuatu's position, and ultimately, the UK did return the territory.

There was a flurry of attention around that win, and it may have sparked some bold thinking. Not long after, a group of budding lawyers paid Minister Regenvanu a visit. During a class at the University of the South Pacific, the students had brainstormed ways to advance climate justice. Frustrated by the limits of law and legislation within their island nations, they came up with an idea that would bind all countries to act toward 1.5 degrees Celsius: Why not seek an ICJ opinion on climate change?

Minister Regenvanu was in. So were Julian Aguon and Margaretha Wewerinke-Singh of Pacific-based Blue Ocean Law. They began to build a counsel team. Given the Chagos work, Jen was an obvious addition, advising on diplomatic strategy—how to get the United Nations to greenlight the climate case—and legal strategy—what questions they wanted the court to opine on. In parallel, the now-formalized Pacific Island Students Fighting Climate Change (PISFCC) would lead an advocacy campaign "to bring the world's biggest problem to the world's highest court."

People thought it was politically impossible to get this case before the court. A majority of UN member states would have to vote in favor of it, and surely the big emitters would block that. But over years of effort, a coalition of countries came together. Not only did it turn out to be possible, the UN sent the case to the ICJ by consensus. *It was wild.*

It's a testament to the power of idealism and a shared dream. The students were not hampered by the cynicism of diplomats or academics. They saw clearly that the court should rule on this—of course it should—and they had the audacity to demand it.

The court would take up two questions: One, what obligations do countries have to take action and protect the climate? Two, what are the consequences if they don't, or fall short, and cause harm? Jen would act for Vanuatu and the Marshall Islands, to help shape their answers to those questions.

Proceedings took place in December 2024 in The Hague, Netherlands. On the first day, the president of the PISFCC, Cynthia Houniuhi of the Solomon Islands, addressed the panel of fifteen judges: "Land is our mother, a living, timeless plane where generations past, present, and future converge, interconnected and sustained in an unbroken cycle of life." It's a matter of intergenerational inequity, she explained, for that land to be swallowed by rising seas.

Cynthia and I walked to court together that morning, and it struck me. She was a law student who had become completely focused on this campaign. She didn't yet have time to begin practicing law in the Solomon Islands, but here she was about to address the world's highest court. And she set the tone in so many ways. It was clear that these proceedings would be different.

The judges heard ninety-six government submissions of half an hour each, so half an hour of different countries from around the world presenting what climate change means for them. Some, like the Marshall Islands, brought science—models of projected sea-level rise showing how climate change will steal their land. Many brought stories about the sheer magnitude of damage. Several representatives teared up as they shared the effects on their people, their cultural legacy, their economic well-being. It was heartbreaking. You could feel the weight of the court's responsibility.

The legal team, activists, and aligned countries were all staying at the same hotel. Each night, they would strategize and practice submissions.

We sat in rooms together, talking about who's coming up next, how can we support them, how can we make sure the arguments best support Vanuatu's position? There was ongoing consensus-building, teamwork, camaraderie. That's how we approached the law and the little things. The attorney general of Fiji forgot to bring his wig and gown, so I loaned him mine for his address to the court.

Over the course of the proceedings, you could feel the global solidarity. The world was taking on the big polluting powers. It was so obvious that the US, China, Russia, Saudi Arabia—they were out of step with the rest of the world. And it was so transparent why: Economic self-interest was driving these terrible legal positions. The aim was to avoid accountability.

On the final day, the PISFCC addressed the court again, this time represented by Vishal Prasad of Fiji. He opened by referencing the traditions of wayfinding: "In the Pacific, we have always looked to the stars." This is more than a method of navigating the ocean, he explained; wayfinding is about a relationship between "those who came before with those who will follow."

The navigation in this moment is worldwide and existential. Prasad went on: "Today the world needs wayfinders, those who can guide us towards a path that protects our homes, upholds our rights, and preserves our dignity." He called on the court to use international law as "a compass for justice and accountability."

And the court did. In July 2025, everyone was back in The Hague to receive the ICJ's opinion.

We were nervous. We didn't know if the court would go as far as we wanted them to go. The court could have wriggled out of it or aligned with the powerful states. As they read out the unanimous judgment, we kept looking at each other, like, they went there! They went there too! They did everything we wanted them to do. They even said it's a potentially unlawful act to subsidize fossil fuels.

At the hotel, there was a party. And the dreaming began anew—what this could mean for climate advocacy, international negotiations, litigation against polluters, reparations for harm, shifting the cultural conversation, and phasing out fossil fuels. The spillover effects will be many and great. Chances are, Jen will be lawyering toward some of those possibilities.

The court also acknowledged that international law—their sphere of action and mine—"has an important but ultimately limited role" in resolving the climate crisis. It will take "all fields of human knowledge" and "human will and wisdom . . . in order to secure a future for ourselves and those who are yet to come." I'm using the skill set I have and the tool of the law to shape and motivate political action.

This "impossible" legal victory started with twenty-seven students in a classroom in Vanuatu. Imagine what we could achieve if we all shared their ambition?

The compass of science has long pointed to 1.5 degrees Celsius, as has the compass of advocacy. Now, the compass of law has joined them. Even as hopes to stay below that threshold wane, these compasses are directing the world toward a future that is actually livable for all. May our many arrows magnetize to the far shore of that vision and draw us ever closer.*

*Jen's story is rewoven from a conversation we had, as well as other material. Hear more from her, as well as Cynthia and Vishal, at climatewayfinding.earth.

Playlist

Songs to see feelingly.

"Bottom to the Top"—Joan Armatrading
"Oh Wide World"—Mon Rovîa
"The River"—AURORA

Tune in at climatewayfinding.earth.

Deepen

Tap into your vision and imaginative wisdom through guided meditation. Visualization can help deepen and expand your ways of knowing possible paths forward. (*~15 minutes*)

What you'll need: a quiet space to be in and the audio meditation, found at climatewayfinding.earth. (If you aren't able to go online and listen to the meditation, you can use the printed version here.)

Three suggestions before you begin:

+ Invite any grumbling or gravity to depart, and welcome in a sense of play and lightness.
+ Embrace any bits of imagining that come, however funny, fuzzy, or far-fetched.
+ Freewrite about your vision immediately after the meditation. (There are no prompts for that, just capture anything you've imagined.)

Vision Meditation

If you'd like to offer this meditation to others, or ask someone to read it for you, here it is in print. Go slowly, allow pauses, and, if possible, model the breathwork.

Begin by coming to a comfortable position for your body. You might want to be more upright and seated; you might want to lie down. Whatever will help you really tune in. Take a moment to get there.

Allow your eyes to gently close or soften your gaze. As you settle in, breathe slowly and deeply, savoring the inhale . . . and savoring the exhale . . .

If it feels good, you might put one hand on your heart and one on your lower belly—connecting physically to those spaces from which wisdom arises.

Take another generous breath in . . . and out . . . As you breathe, notice again the firmness of Earth beneath you, the way Earth holds you. Right now. And always.

(*If in a group*) Feel the presence of each person in this circle, here together, in this moment.

Allow your imagination to float forward to three years from now—a time when you are immersed in climate engagement that feels deeply soulful.

The day is breaking, and you are just waking up. *What do you notice about the morning? Where are you? What calls for your attention as you start the day?*

This day will be spent in meaningful ways. *What excites you about the hours ahead? What's on your list of priorities for the day? Are any larger projects in motion?*

As you move through the day . . . *Who do you encounter? Are you collaborating with anyone? How does that collaboration feel?*

What skills are you using? What are you learning? What are you teaching?

What is giving you a sense of authentic power? What is giving you a sense of deep joy?

There is a challenge in your day, and you are well-equipped to address it. *What resources do you notice, internally or externally?*

There is cause for celebration on this day. *What are you celebrating? And how do you celebrate it?*

It is evening, and you are winding down to sleep for the night. *What are you feeling grateful for? What are you feeling nourished by?*

Notice how this day of meaningful climate engagement feels in your body. Take a generous breath in, savoring the goodness of your day . . . and take a generous breath out, settling into a sense of fulfillment.

Staying with these sensations, slowly open or refocus your eyes and freewrite about your vision of future climate engagement.

Map

Create your own Climate Compass—synthesizing insights from the *Climate Wayfinding* journey—to guide you as you go forward from here. (*20–25 minutes*)

What you'll need: your freewriting, maps, and notes from previous chapters; a pen; your journal and/or a printout of the Climate Compass, found at climatewayfinding.earth.

+ Begin by reading back through everything you've written, mapped, and discovered, from the beginning of *Climate Wayfinding* to now.
+ With all of that in mind and heart, begin to distill and summarize insights. Draft each of the six segments of the compass. (If you get stuck, freewrite.)
+ Adapt your "canvas" as needed. If the printout feels confining, use a big sheet of paper or write in your journal, and then organize it into your compass. If a compass doesn't feel right to you, make a collage, poem, patchwork sail, coat of many colors—some synthesis that suits.

The Climate Compass is designed to be a living tool, so continued reflection and refinement is a wonderful thing. This counsel feels especially apt—from adrienne maree brown in *Holding Change*:

> *Release perfection. The idea of iteration is that we are repeating—not failing, but practicing and learning. And with each repetition there are things to learn, notice, grow from. Love the body that does the practice, love the imbalance that yields to balance, love the shaking muscles that grow strong, love the fumbling tongue that becomes fluent, love the chaos that finds an organic and exciting center. Let love show up and change the possibilities.*

Journal

Reflecting on your vision and compass, begin to freewrite. (If you skipped the Vision Meditation and/or mapping your Climate Compass, sweep back through, as these are of the essence.)

> What feels especially life-giving in all that you've envisioned? What direction(s) could your deep, sustained, courageous climate engagement take from here?

> If this is a moment of telling yourself the truth, what else do you need to write down—and ensure your future self knows?

Step Out

To make the Climate Compass a useful living tool, you'll consciously deepen a relationship with it. That could take a few different forms. (*10–20 minutes*)

+ Print and post your compass somewhere that feels present yet sufficiently private.
+ Record a voice memo of yourself reading through your compass. Listen to it periodically to stay in relationship with the direction that is growing inside you.
+ Share your compass with someone in your life. Ask that they witness your sharing in generous listening mode and offer affirmations.

Return to your compass often to keep it living and growing inside you. Evolve it as needed. It may be especially useful when you're facing a decision. Reflect: *Does a given path align with the soulful sense of direction you set down?*

Gather

This is a 75-90 minute experience.

Reminder: Everyone should bring their Climate Compasses.

Tunes (*~10 minutes*)

Play songs from the playlist as your group gathers.

Poem (*~5 minutes*)

Read "Dead Stars" aloud. Then, invite a second reading.

Opening Circle (*~5 minutes*)

Begin an opening circle with this prompt—modeling first, then popcorning.

> Share a body posture or movement that evokes the idea of vision or "speaks" to some aspect of the vision you're holding. When someone does their posture or movement, we'll all "echo" it with our own bodies.

Reflection (*~15 minutes*)

Ask the group to look back at their drafts of the Climate Compass and freewriting from this chapter—and jot down any additional thoughts. It might be useful to further refine elements of the compass. (Allow a bit more time for this than normal.)

> What feels especially life-giving in all that you've envisioned? What direction(s) could your deep, sustained, courageous climate engagement take from here?
>
> If this is a moment of telling yourself the truth, what else do you need to write down—and ensure your future self knows?

Discussion in Trios *(~20 minutes)*

Invite everyone to form trios and share their visions and/or compasses. As ever, keep time and let people know when to switch. (Again, allow a bit more time for this than normal, with rounds of about 5 minutes.)

Discussion as a Group *(10–20 minutes)*

Return to a whole group, and invite people to share from their trios. If useful, bring in these prompts.

> How was the experience of visioning? Of drafting the compass?
>
> Did anything in this chapter rouse your memory or quicken your imagination?

Closing Circle *(5–10 minutes)*

End with a closing circle using this prompt. Invite someone to start; then popcorn.

> Share one key word or phrase from your compass. As you speak it, we'll hold it together.

Closing Quote *(1 2 minutes)*

Read an inspiring passage from this chapter.

8

Journeying Onward

Set your compass. May each step from here help to heal our shared home.

I have come to Taos, New Mexico, riding the light of long, nearly solstice days. The high desert has been spared fire so far this year. It offers a welcome respite from the already-thick humidity of Georgia in June.

The trip's impetus is to work with a writing collaborator—gloriously, for the first time, in person—but mostly I am here to shift the frame. I need a week to set down the burdens of home.

Two weeks before my arrival here, my father died, just shy of his seventy-fifth birthday. Alzheimer's had taken him apart, block by block, over a decade. The disease sets up shop in the seahorse of the hippocampus; one of the first symptoms is losing your way. My dad lost his ability to follow well-grooved routes—*Boo, I'm . . . I'm not sure which way to turn*—and ultimately any amount of ambulation. He lost his stories, then later almost all of his words. Those losses were especially gutting for a man who built a vocation around them as a sports journalist.

Being with him at the threshold of life's end, and walking with him as long as I could, has shifted something in me. The fathomlessness of this whole experience—to be a person, in a body, for some uncertain number of years on this Earth, then suddenly gone—it feels altogether infinite. I find myself wading through a raft of new wonderings about liminality and the journey on from here.

In the aftermath of loss, existential questions often sweep in. Questions about what makes a good life and matters most. Questions about impermanence and the twin truths of intrinsic possibility and inevitable ends. The probing, clarifying nature of loss can, I think, be one of its gifts. And the questions seem to have a way of tagging along, even as I cross the Mississippi River and the Great Plains to reach the Southern Rockies.

Of all the wonderings that I've brought with me to Taos, the ones that feel most weighty and most insistent are about home.

> *Will the meaning of home shift as family roots that have held me loosen? Is this a moment to step back and think anew about where and how to live?*
>
> *Amid so much disruption in the world, how do we all find or feel at home?*

Home, the dictionary warrants, is about where we reside. The word's typical use suggests something fairly fixed, even permanent—a place we remain, or, if we leave, to which we return. *Homecoming* is ritual. Animals *home*, returning by instinct to their territory. To be *at home* is to be at ease, at peace.

> How does the idea of *home* reverberate for you?

In the thick of the climate crisis and its abutting troubles, however, places we call home today may become unrecognizable or be lost completely. Some—like Bayou Liberty, Louisiana, where Colette Pichon Battle lives—almost certainly will. Intellectually, I comprehend the acute wildfire risks in California, yet I was still stunned when cousins in Los Angeles lost their home to one. *Which way to turn* takes on another, awful, meaning.

Wayfinding, in very literal ways, will be part of climate futures ahead, and it will not always be a choice. At times, it will be the catapulting outcome of turmoil and displacement, which are all the more difficult in a world of borders and barriers, of battle lines around who belongs. How many hundreds of millions of people may need to migrate has everything to do with how hot it gets and how high the oceans climb—whether temperatures are survivable and land is still land.

Home is in the crosshairs. The need to *re*-home is sure to be common. The questions of *where home is, what it means, how to make it*—they feel weighty and urgent for many. I'm beginning to accept that these burdens cannot really be set down. Maybe the shifting sands, both underfoot and within, are calling us to embrace the interplay of rooting and roaming that is so core to humankind. To be an Earthling is to be a denizen of change.

A core credo has accompanied us throughout *Climate Wayfinding*: to welcome the questions we are holding, however daunting they may feel, and to honor them as both a means of orientation and a source of sustenance. In my experience, the arrival of questions, or their return at a different pitch, is often a first sign of the crossroads. They are a gesture of the juncture—that place we come to again and again.

In a beautiful conversation with journalist Krista Tippett, climate leader Christiana Figueres shares the inquiries buzzing within her own climate journey: "During the time that I'm here, which is a blink in the history of 4.5 billion years of this planet . . . what kind of a blink do [I] want to be? Who do I want to be? How do I want to turn up in the world?" We cannot expect any guaranteed outcomes from our answers, she counsels, but they nonetheless "influence the direction that we move in."

Through the passage of *Climate Wayfinding*, we have taken time to be with such questions, whether long-pondered or newly sprung, and pause in the neither-here-nor-there of our intersections and gateways. If I'm honest, that pause is a place I like to linger—out upon the depths of wondering. I like the breathing space of a standstill, even if its cause is unnerving. I like how the throughgoing nature of querying and exploration absorbs me . . . awakens me . . . enlarges me.

At the same time, crossroads and thresholds entice us to set off. The answers we whittle begin to propose a foot- or flight path. Having been in a liminal space, especially with time and support to do so well, can stir a readiness. Like train wheels longing for a track or sails eager to meet the breeze, we want to step across, take a turn, and set out on the next leg of the journey that is drawing us on—even if it doesn't entail physical departure at all.

In many ways, continuation is the essence of *Climate Wayfinding*. The word *wayfinding* is a noun, but its ending—*ing*—denotes motion. *Found* is final; *finding* is ongoing. *Way* offers us insight too. It is a route or path for traveling, a direction of movement, a course of action. It is also a method or manner of doing something, as in, our characteristic modes, patterns, and behavior. Across these meanings, there is a sense of taking steps and going on.

Climate Wayfinding, then, cannot bring us to a final terminus or clear conclusion, and it shouldn't try to. This chapter of life is reminding me, viscerally, that the act of finding our way will only continue to open. Seismic shifts, in both our inner terrain and the wider world, will upend yet more maps and compel new cartography. As humans on a planet that we hope to heal, *onward* will likely be a never-ending story—one in which we discover, again and again, what is essential *now*. The path carries on, on, on beyond the page.

The trick, I think, is to head toward the future without leaving the liminal behind. As we try to tilt the world toward ecological and social revival, we will need to continue looking inward with care and outward with curiosity *alongside* looking forward with courage. Three-way perception is truly vital—from the Latin *vita*, meaning "life." We might imagine our sensibilities of questioning, listening, sensing, and connecting as precious paddles that we carry with us as we navigate a stormy age.

Indeed, there is much from the *Climate Wayfinding* experience that can sustain us: old stories and emotional heaviness released; insights revealed and new stories gained; practices learned and capacities developed; relationships formed or grown; invigoration sparked or revived. We may feel more equipped with healthy and wise ways of being. When we find ourselves again at a crossroads or threshold, perhaps in a moment of feeling utterly stuck, we can return to the ethos and modes of *Climate Wayfinding*—and to the things we wrote down when telling ourselves the truth.

Navigation needs nourishment, I remind myself as I get the lay of new land during my own present journey. So, *don't rush.* Sometimes we need to go inward to be able to move forward, and there is always something sage within, waiting to surface. We can make space for it to do so through writing, sketching, contemplative practice, embodied expression, and deep dialogue. In parallel, we can tune to what the world beyond our beings has to say—especially in its quieter corners—to the human *becomings* dwelling here.

> You embarked on this experience with key questions in hand. What questions are with you now?

My abode for the week in Taos, my "home" for the sojourn, is a couple miles southeast of the historic downtown, just where the Rio Fernando de Taos descends from the mountains and arrives at a thirsty valley. The one-room adobe structure was built in the early 1950s—a dance studio for María Benítez, who was an astonishing flamenco performer and choreographer. She practiced in this space for a decade, I'm told, before entering the international scene. I imagine innumerable taps and stomps landing upon these wooden floors. There is a sturdiness and a rhythm to the place.

Flamenco was born more than five thousand miles east—at nearly the same latitude—in Andalucía in the south of Spain. Roma migrants from northwest India arrived in the region in the fifteenth century. There, Romani cultural traditions mixed with those of Muslims and Sephardic Jews. Flamenco music emerged, and then dance—creations of marginalized people, persecuted by Spain's Catholic elite. Art born of keeping on. Art born of wayfinding.

"Gutsy" and "earthy"—that was the description Benítez once offered for how she wanted to be and move. Ballet, she discovered early in her dance career, was not right for her, either physically or in temperament. In flamenco, there is passion and power transmitted through the body but sourced from within. It often has a transmuting quality: pain, devastation, anger, or longing can metamorphose through movement into determination and defiance, joy and aliveness. Spiritual transcendence can arise from its emotive smolder or its percussive fire.

Flamenco is adept at creating space for the variegated emotional range of human life, not only within the artists but in the audience too. The performance is itself a threshold, inclining us toward something new or reawakened within. Through movement we can heal, even if mending every fissure isn't possible.

To mend every fissure is the conventional sense of the word *healing*. But it is complicated by the realities of Earth, which has always changed and is changing so much, so fast in our lifetimes. The same could be said of life at a small scale. Some things can be repaired or put to right; some renewal and equilibrium remain possible. Elsewhere, a cure is elusive; the only option is care and then hospice.

Here on our home planet, we face something much more complex than either a return to full health or a terminal diagnosis. We may be hospicing a great deal, but we are also welcoming and nurturing new life at the same time.

Alaska issues its first-ever heat advisories, and the Klamath River, undammed, runs free from Oregon to California. Gas turbines powering the world's largest supercomputer foul the air of southwest Memphis, Tennessee, and a conservation group buys, and blocks, a planned mining site that threatened to imperil Georgia's Okefenokee Swamp. There is, as there always has been, simultaneity of tomb and womb. We exist, on occasion, at their portals, but mostly in the interstitial country between them.

Healing, to my mind, is not about obtaining a definitive, unfailing state of good health. To heal is to stitch up what we can, as we grow greater range—expanding our capacity to find a wholeness of paradox and fragments. Sometimes, healing is about fixing. Always, it is about including the wound. The task of this time is to embrace the entirety of what is happening on our planet—not in every specific detail, which would anyway be impossible, but in the knotty essence of it.

> Imagine an expansive view of *healing*. What would you fold in to make it more helpful and whole?

In the loam of my psyche, a new question is sprouting:

How do I make of myself a home that is expansive enough to hold all of it—the rootedness and roaming, the devastation and defiance, the knowingness and mystery?

To be home, I'm coming to see, is to say *yes*—*yes* to the needs quaking across our planet, *yes* to this time of trouble and transformation, *yes* to the persevering possibilities of life—even if I say *yes* with tears in my eyes or a howl in my chest. To be home is to honor that there is so much beyond our choosing, *and yet* we also have choice. All is hallowed ground.

To say *yes*, in an embodied way, is to move into action. Home, after all, needs makers—those who mold the places we live into shelter in the fullest sense.

As we know, the field of *doing* that could assist in planetary homemaking and healing is vast. Byways abound.

We have shaped a Climate Compass to help us orient within that field, honing what is primary, filtering possibilities, and making the path forward more clear. Alongside that tool, another framework can accompany us. It's a triptych of the realms in which our *yes* makes a move: *personal life, professional life,* and *public life.**

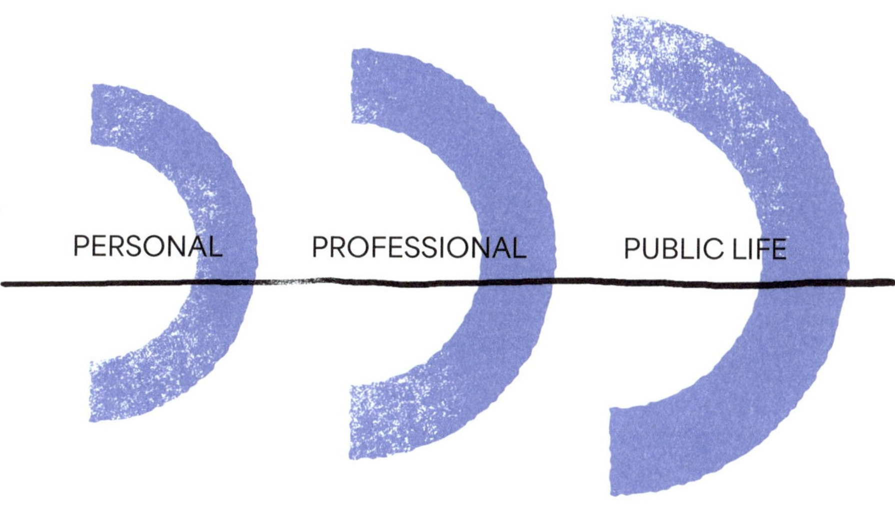

3 Spheres of Action—Personal, Professional, and Public Life

*Dr. Leah Stokes and I adopted a similar framework for a three-part podcast series on *A Matter of Degrees*. In it, we explore the personal, professional, and political dimensions of that essential question: *What can I do?* Find those episodes at climatewayfinding .earth.

Crucially, these 3 Spheres of Action are not discrete or impermeable, but side-by-side and connected. Yet seeing them with some differentiation can help us comprehend the landscape and bring structure to the actions we wish to take. Rather than picking just one, we have an opportunity for motion in all three at any given time. (See "Sample Actions in the 3 Spheres.")

PERSONAL LIFE: Here, we find the things that are closest in, dotting our day-to-day. The buildings we inhabit. The energy-swigging machines around us and their alternatives. The institutions that invest our financial resources. The things we eat. The ways we travel or choose not to travel. What we buy or don't. What we do to reduce waste or bring its worth back around. The life skills we cultivate and lend. How we care for climate-related well-being—our own and others'. How we engage with our families, friends, neighbors, and how we build community. How we engage with our immediate more-than-human world. This sphere of action can keep us feeling connected with our values daily. With so much we can do, we may wish to focus on actions that create meaningful, durable change.

PROFESSIONAL LIFE: Here, we encounter the intersection of climate and our work-related pursuits, including education. The climate-related jobs, gigs, or internships we hold or seek out. The coursework or degree programs we undertake to learn and grow our capacities. The more informal learning we do. How we mentor and are mentored. The ways we make our current work into a site of climate engagement. The openings we identify, and pursue, for change in our institutions. How we experiment and innovate with new methods, initiatives, enterprises, and creations. This sphere of action is teeming with opportunities, especially given that so many of our waking hours, and so much of our energy, are spent here. We may find paid work in climate engagement and the riches of collaboration too.

PUBLIC LIFE: Here, we come into our interfaces with civic life and broader society. The groups we are a part of. How we shift the cultural and spiritual fabric around us. How we engage in spaces where decisions are made for the whole: community meetings to city councils to congress or parliament; local parks to transit authorities to regional food systems. What we do to uplift democratic systems and climate leadership in public office. What we do to advance change through advocacy, accountability, and dissent. The ways we stand up, or sit in, to protect water, land, and community. The ways we pioneer new laws, economic models, or social systems. This sphere of action is where the *we* is pivotal. We may discover that the realms of personal and professional life ripple into this one, with prospects for collective waves.

When it comes to laying plans to enact our *yes*, it can be tempting to take on entirely too much. Our fields of action, then, become too many, and dispersion scatters our efforts, dimming their impact. I've certainly overdone *yes*, often as an attempt to escape the creeping concern that my contributions are ever so tiny in the nearly infinite context of our planetary trouble. Sometimes *no* can be a co-conspirator to a more resonant *yes*. Counterintuitively, we may seed a prolific flowering by looking to the coming season and committing to one, two, or maybe three action steps for each of these areas. We take them and let momentum build for another step and then another—perhaps bigger, perhaps bolder. *What can I do . . . next?* turns out to be a very helpful version of that essential question.

We might also give ourselves the grace of releasing the idea that an embodied *yes* only looks a certain way and must have a huge or immediate impact. At worst, the anxiety we may hold about what counts as "worthy" can keep us from acting at all. At best, a more expansive view of *doing* might send forth the most spectacular shoots. Imagine that all we do has alchemic potential. Imagine that there is something sacred in every act.

> Where might these three spheres be inviting you to go? What sacred doing might you say *yes* to?

In 2009, a group of renowned Earth system scientists—led by Dr. Johan Rockström and including one of my doctoral supervisors, Dr. Diana Liverman, and a future colleague, Dr. Jonathan Foley—introduced a framework known as the *planetary boundaries*. They are nine critical processes that, together, underpin Earth's stability and resilience. Without them, we forfeit a "safe operating space for humanity." There is now an annual Planetary Health Check. As of 2025, seven of nine boundaries are outstripped.

The simple math of what needs to be done, to slow the breaching and bring us back into some kind of balance, is mind boggling. I close my eyes and feel a throbbing at my brow. We will need the kinetic power of enormous, elegant wind turbines and solar power from the transfiguration of light. We will need the legislative power of governments and the informational power of media. We will need people power as abiding and tenacious as dandelions.*

But I think we will need power from yet another dimension, if we want to do what needs doing and get where we need to go. The planet has boundaries, yet circuits with something so very boundless too.

Almost exactly forty-eight hours after my father died, I woke up in the night, quick and lucid. Right above my chest, I felt pulsing, porous energy—like sequins catching light, only a sensation rather than something seen. It was nothing I've ever experienced before, yet somehow so familiar. It was, as best as I can describe, the feeling of my dad's sheer and absolute love, which he gave with such ease.

The rawness and great enigma of death may let us touch into the reality of love that is so much bigger than a body. It flows in a kind of malleable, connective latticework—a gauzy webbing that is as present, suffusing, and necessary as air, and equally invisible. I cannot possibly prove it, but I know we are all caught in Earth's net of unstinting, unequivocal love. Our deepest belonging sources from it. Our deepest learning resides there. And our deepest invitation is to participate in the exchange.

*In recent years, these flowers have become a global symbol of climate justice, thanks to Project Dandelion—a women-led campaign catalyzed by Mary Robinson, global elder, advocate, and former president of Ireland. Dandelions are widespread around the world, often thriving in seemingly inhospitable places.

I long for climate healing motivated not from wanting to *fix* something but from wanting to *love* something.

> *What might open up if love quickens the steps of our doing and engagement?*
>
> *If we let love run riot—all weedy and untamed—what dogged blossoming might come into view?*

To be sure, I am not saying that *feeling* love will be enough. It won't be. We are wise, in all facets of life, to be skeptical of love that is named but not enacted. But enactment of projects or policies or protests, of sharp strategies and sincere missions and movement building—it simply cannot do or be enough without that spirit, glowing and glittering at the core. In the absence of it, we can wind up replicating the very paradigms and systems we are trying so hard to repattern.

Beneath or beyond my throbbing mind, I notice a quivering desire. I want to be part of the humming heartbeat of Earth as long as I am able. When my own inner rhythm stops, I want to know that I was a sequin on the golden skirt of life's dance. I played my part in the shimmy and the shimmer, even as the world lurched.

> Have you been witness to love springing up or streaming in—shifting what could be?

My last day in Taos lands on the solstice. There is writing still to do, but the body yearns to be one with water. We shut laptops and drive down to a park in the Rio Grande Gorge.

Rivers are known for sculpting the landscape in spectacular ways, but in the case of the Rio Grande, the landscape has also uniquely shaped its course. The river flows through a continental rift zone—an area where the Earth's outer skin is being stretched and thinned as tectonic plates move away from each other. This particular rift has been opening, ever so slowly, for some twenty-five million years. In a sense, it's given the river a place to run.

At the end of a dusty road, we come to a confluence. The slim Rio Pueblo de Taos flows from its headwaters at Blue Lake—the sacred heart of the Taos Pueblo homelands—and the even slimmer Rio Fernando de Taos slips in along the way. Here, the waters spill over tan-gray rocks and into the big river, the fleet-footed Rio Grande. Watching their cascade, I ponder whether they will endure the long, parched west flank of Texas and reach the Gulf of Mexico.

Just above the confluence, there is the perfect shallow pool for sitting in cool waters at the year's height of light. All around us, babble and whoosh, glint and gleam. The pool's surface is a starry spectacle—the playful collaboration of sun and current.

Along the rivers' shared edge, horsetail is thriving. *Equisetum*, I learn, is the only surviving genus of a primordial lineage, profuse during the Carboniferous period (roughly 359–299 million years ago). These reedy plants are a miniature descendant of tree-sized ancestors, the remains of which are abundant in deposits of coal. I learn, too, that horsetail is known for its capacity to strengthen connective tissues, like skin, bones, ligaments. It can help heal fractures and mend wounds.

On this bright day, in this one pluck of the song of life, there is so much *everything*.

I think of the nodes that we each are—connected through time and space, all contributing their own quiver *to* Earth, *as* Earth. There are moments when we recognize ourselves as nodes grown stronger with possibility. My hope is that this is such a moment for you.

At soul points in our navigation, we hear more deeply the wisdom within us. We are more attuned to what it would mean to express life force in our days.

We feel more equipped to stride toward that, sensing the path as we go. It's as if we're able to draw a new line in the constellation.

We are still under the same sky, but we are no longer on the same ground as where we began the *Climate Wayfinding* journey. Can you feel a fresh firmness beneath you? Can you feel a magnetic pull for the gifts you are here—in this time, in this place—to bring?

You, with your tender heart and keen mind and profoundly capable hands, are a wonder. You are remembering and conjuring; you are heirloom and maker; you are an essential part of Earth's abundance. Let us marvel at what might yet unfurl. Let us find our way in the sureness of knowing love is the mystery and the gravity that holds all of life together.

The way forward? I trust that we know it by heart. That way is calling us home.

Summons

Aurora Levins Morales

Last night I dreamed
ten thousand grandmothers
from the twelve hundred corners of the earth
walked out into the gap
one breath deep
between the bullet and the flesh
between the bomb and the family.

They told me we cannot wait for governments.
There are no peacekeepers boarding planes.
There are no leaders who dare to say
every life is precious, so it will have to be us.

They said we will cup our hands around each heart.
We will sing the earth's song, the song of water,
a song so beautiful that vengeance will turn to weeping,
the mourners will embrace, and grief replace
every impulse toward harm.

Ten thousand is not enough, they said,
so, we have sent this dream, like a flock of doves
into the sleep of the world. Wake up. Put on your shoes.

You who are reading this, I am bringing bandages
and a bag of scented guavas from my trees. I think
I remember the tune. Meet me at the corner.
Let's go.

✡ Lighting the Way

with Sherri Mitchell (Monroe, Maine)

We met Sherri Mitchell (Weh'na Ha'mu Kwasset) at the beginning of this journey, when we greeted the liminal and learned her metaphor for the space between: the air held within a drum, where vibration kindles sound. As we turn our last pages and go forward from here, it seems right to meet her again. Sherri is a teacher and author who not only brings us to the waters of wisdom but also helps us add our unique tributaries to the flow.

Sherri Mitchell was born Penawahpskek and grew up on the tribe's sovereign territory, located within what is now called the state of Maine. Wabanaki means "People of the Dawnland," and the Penobscot Nation is one member of that tribal alliance, rooted in the far east of Turtle Island.

We are the keepers of the eastern gate and have been for more than ten thousand years. Our homelands are located along various waterways in Maine and the Canadian Maritimes. It's a riverine culture. I learned how to paddle a canoe on the Penobscot River before I ever learned how to ride a bike.

That land and those waters are where I first learned of my place in creation. We are the collection of individual notes in one continuous song—the song that sang all life into being. If we listen closely, we can hear this creation song echoing in our bones. Sourced from the universal frequency, each individual soul carries its own unique vibration. The symphony we create all together brings into form the reality that we see before us. To shift the reality that we live in, we have to shift our vibration, harmonizing with the rest of creation.

When Sherri was young and imagining her note in the song of life, she wanted to be a writer. But writers, she was told, don't make money. (Economics of publishing aside, she eventually came back around to that calling with fire.) After college, Sherri worked for a number of years as a civil rights educator, before somewhat unexpectedly being recruited to law school. It felt like the path opened up before her—a path that took her to Arizona and beyond.

In 2011, while living in Tucson, she founded the Land Peace Foundation, a nonprofit initially focused on protecting water and sacred sites. Over time, it came to focus on Indigenous and environmental education, and at that point, her kids—grown and both living back in the Dawnland—got wise. This was work their mom could do from anywhere, and they wished her home.

I didn't think I would ever come back to Maine. But my daughter and son were persistent. After driving five days across the country, I remember coming around a corner and I saw the Penobscot River. I burst into tears—ugly crying. When I was a kid, we were forbidden from going in that water because it was so polluted. My tribe spent thirty years cleaning it up, and I saw kids swimming in the river for the first time. I thought, OK, this is where I need to be.

Sherri had plans for a "quiet, normal life" and lots of time to be alone. But her legal work started taking her around the state, giving presentations to support the ongoing protection of the Penobscot River. As time went on, crowds got bigger and bigger still, while the topics she was speaking on broadened.

I showed up in this very tiny town, where they had asked me to speak, and it was a total traffic jam. I finally recognized one of the coordinators, wearing an orange vest. "Tony, what the heck is going on?" I asked him. "This is for you," he said. "They've got a parking spot saved over there." I went around the back of the building and cried. And then, reluctantly, I surrendered to a different life. That simplified life was never my path.

Even with that clarity, Sherri realizes the field of possible doing is vast. Two things guide her as she finds her way within that meandering green that goes on virtually forever.

First, I grapple with wanting to contribute. There's this deep soul movement, drawing me to be of service at this time. I have a guide that's come to me in my dreams for many years. In one dream I asked him, "Am I doing the right things? Is it enough? Am I wasting my time?" On and on and on.

After I was done, there was a long pause. He said to me, "Tell me, what are you working so hard for, if you have no joy in your life? It doesn't matter whether you are standing in front of a crowd of fifty thousand or serving sandwiches behind a deli counter, your role is to be the most authentic version of yourself and to bring your full vibrational presence to the harmonics of this planet." I think about that a lot: What is my authentic move right now?

Second, I recognize that I want to be effective, whatever I do. I think about my light as a laser. If that laser becomes too diffused, because it's aiming in too many directions, it becomes powerless. But if I know where I want to direct my energy and stay focused on that, I can have a powerful impact.

Sherri first shared with me the story of her guide's counsel as we drove a tiny red rental car—an overgrown roller skate, really—through Joshua Tree National Park. I was racked with general climate despair at the time, as well as more personal despair about the "best," "right," and "enough" work to do. She told me that story, and then another.

In our mythology, this is the fourth time humans have been here. This is the fourth iteration of us existing in this form, in this learning environment. We are still trying to reach a fifth world. We may not get there, but however far we progress, we will come back and try again. We will pick up where human consciousness is now and then spiral forward.

It gives me a sense of measure. I don't feel frantic. I'm going to do the very best I can, while I'm here, to shift hearts and minds and grow consciousness. I feel that's my job. And I'm going to do it by focusing on things that are in relevance to the life I was born into.

For my part, the myth brings me to a similar place as W. S. Merwin's poetic oath: "On the last day of the world / I would want to plant a tree." It relieves me from the biggest what-if questions, like, *What if human life doesn't go on?* It gives me reassurance that what you and I offer, from the depths of ourselves, cannot be wasted. It gives me resolve that whatever ground we gain will be beneficial in far-reaching ways.

It also reminds me: Stay rooted in the truth of where we are as a species.

We will not magically arrive at the next place, and we are not on the verge of nirvana. The next leading edge is to let go of that fantasy and recognize that the distance we have to travel cannot be bypassed. We actually have to walk the whole distance, and we have to learn to walk, shoulder to shoulder, out of the desert together.

We want the evolutionary leap. But humanity is at the same stage of development as Hazel, my granddaughter. We're a young species, and that's OK. This beautiful gift of learning that we have, on this planet, is *the deal. We have to go through the steps of emotional and spiritual development, while keeping our feet connected to the Earth. One step at a time.*

I asked Sherri if she had a wish or a blessing for all of us journeying onward from here.

Trust the rhythm that is in your bones. We are so busy trying to think about who we're supposed to be and what we're supposed to do that we forget to explore who we are in truth. *If we can find a pathway to that truth within us—and can bring that, fully embodied, into the world—it's medicine.*

Each of us was born for this time, on purpose. We each have something that can help heal and grow the consciousness of the entire planet right now. Our most powerful activism is to become fully ourselves in deep connection to our heart-based wisdom and divine knowing—and to harmonize our being with the rest of life.

The first step on that journey is coming to know the truth of who you are and allowing that to shine unabashedly.

Yes, may we shine unabashedly. May we gleam as the very people we were born to be—in these bodies, in this life, on this Earth—and offer our notes to the song of creation.*

*Sherri's story is rewoven from a conversation we had, as well as other material. Hear more from her at climatewayfinding.earth.

Playlist

Songs for all the paths from here.

"Pachamama (BC Version)"—Beautiful Chorus
"Find Your Way"—Rising Appalachia
"Higher Ground"—Stevie Wonder

Tune in at climatewayfinding.earth.

Map

Return to the 8 Dynamics of Climate Engagement to trace the shifts that have occurred within you. *(5–10 minutes)*

What you'll need: your original 8 Dynamics results; the 8 Dynamics Quiz, found at climatewayfinding.earth. (If you aren't able to go online and use the quiz, capture your answers in your journal.)

+ Take a moment to reflect on each statement and consider what is real for you right now.
+ For each dynamic, mark your response from 1 (*that's not true for me*) to 5 (*that's extremely true for me*).
+ After taking the quiz, save your final "web."
+ Compare where you started this journey and where you find yourself now. Notice where you've grown and where you'd like to grow further.

Journal

I posed these questions in the essay of the first chapter, "Embarking." Having taken the journey of *Climate Wayfinding*, now let your own answers spill onto the page.

> What can I offer to our world?
> Is my soul built for certain contributions?
> Who am I here to become?

And the journey continues. It takes courage to find our way in climate healing. That process is evolutionary, imperfect, potent. Freewrite a message to yourself—one you can return to as you continue on the path.

> What wish or blessing would you like to offer yourself? What words of wisdom or encouragement for journeying onward?

Step Out

It's time for the Climate Compass to begin to guide your next steps from here: *Where would your compass steer you within the 3 Spheres of Action?* Imagine some tangible action steps—gestures of sacred doing or ways you might embody your *yes*—that you could take in the coming season or the next three months. (*15–25 minutes*)

Open a fresh spread in your journal and begin.

+ Look at your compass through the lens of the *personal*. Brainstorm actions you could take. (*2–3 minutes*)
+ Look at your compass through the lens of the *professional*. Brainstorm actions you could take. (*2–3 minutes*)
+ Look at your compass through the lens of *public life*. Brainstorm actions you could take. (*2–3 minutes*)
+ Read over all three lists while sensing into your energy and knowing. Circle or star just one or two priorities in each list—what you most want to commit to in the coming season.

Consider sharing what you intend to do with someone you trust. Ask for any support or accountability you'd find helpful.

And then, take the plunge.

Sample Actions in the 3 Spheres

A comprehensive accounting of climate-healing actions would fill the pages of this book and more. This list is a mere smorgasbord sampling—small, big, inner, outer, one-off, new rhythms, and then some. It's meant to spark your own brilliant ideas for what you might do in the coming season.

Personal Life

+ Buy only new-to-me clothing (except socks and undies).
+ Do walking meditation daily for 10–15 minutes.
+ Explore commute alternatives—walking, biking/scootering, public transit, maybe an EV.
+ Switch to a climate-friendly bank and retirement fund.
+ Organize compost pickup for my community.
+ Do an assessment: Where am I burning fossil fuels at home (e.g., HVAC, water heater, stove, dryer)? How could I replace them and go electric?
+ Prepare for extreme weather: Take a first aid course and organize a go-bag.
+ Write (and publish??) a love letter to the land/ecosystem where I live.

Professional Life

+ Brainstorm ten ways my current job could be a climate job; pick one to move on.
+ Host a climate community-building event at work. (*Make it fun!*)
+ Journal about why I feel stuck in my current role/industry—and how to get unstuck.
+ Sign up for a course to grow knowledge/skills for a professional evolution or transition.

+ Take the leap: Scout and apply for five climate-focused roles that excite me.
+ Look for replicable inspiration: Research the boldest climate initiatives in my sector.
+ Have a dream-and-scheme session on climate action with close colleagues/collaborators.
+ Plan a campaign for my org to shift to fossil fuel–free retirement savings plans.

Public Life

+ Find a statewide climate advocacy group and sign up for action alerts.
+ Research how my utilities are regulated and how to get involved.
+ Do phone banking/door knocking to get out the (climate) vote.
+ Organize a climate grief ritual in one of our city parks.
+ Do a training on running for office; start scheming my policy platform.
+ Volunteer to help elders with emergency preparedness.
+ Attend an upcoming climate-related march/rally—bring friends!
+ Lead a *Climate Wayfinding* group at the public library.

Gather

This is a 75-90 minute experience.

Tunes (*~10 minutes*)

Play songs from the playlist as your group gathers.

Poem (*~5 minutes*)

Read "Summons" aloud. Then, invite a second reading.

Opening Circle (*10–15 minutes*)

Begin an opening circle with this prompt—modeling first, then popcorning.

> Share one way you sense your own "node" has grown stronger with possibility over the course of this experience.

Reflection (*~5 minutes*)

Ask the group to look back at their freewriting from this chapter and the list of actions in personal, professional, and public life—and jot down any additional thoughts.

> What can I offer to our world? Is my soul built for certain contributions? Who am I here to become?
>
> What wish or blessing would you like to offer yourself? What words of wisdom or encouragement for journeying onward?
>
> What are my priorities for the coming season in the 3 Spheres of Action?

Discussion in Trios (*~15 minutes*)

Invite everyone to form trios and share. As ever, keep time and let people know when to switch.

Discussion as a Group (*10–20 minutes*)

Return to a whole group, and focus your discussion on the actions each person intends to take in the coming season. (You might ask them to write/post *one key action* in the space, as a way to honor the movement afoot.)

Depending on the dynamics of your group, you might also explore whether there's any "sacred doing" you want to undertake together. (If you do so, allow a bit more time for that conversation.)

Closing Circle (*10–15 minutes*)

End with a closing circle using this prompt. Invite someone to start; then popcorn.

> Share some words of wisdom you'd like to offer to yourself as you journey onward from here.

(Optional) Linking Hands (*3–5 minutes*)

If it feels right for your group, you might then do a practice of slowly, mindfully linking hands. (Beautiful Chorus's "Pachamama" is a wonderful accompaniment for this exercise.)

Begin by turning to the person to your right, making eye contact, and taking their hand in yours. That person then turns to the next person and takes their hand. On and on, until the whole circle is connected.

Breathe together and feel the strength of your group that will carry forward, even as paths branch out from here.

Closing Quote (*1–2 minutes*)

Read an inspiring passage from this chapter.

I have lost my way many times in this world, only to return to these rounded, shimmering hills and see myself recreated more beautiful than I could ever believe.

Joy Harjo, *Secrets from the Center of the World*

Giving Thanks

The people who have encircled and supported *Climate Wayfinding* are some of the loveliest humans I know. Christian Leahy, you are the writing collaborator of a lifetime. I felt so seen, held, stretched, and honored in the process of creating this book, which has your gifts imprinted on every page. Thank you for helping me reach my depths. Amy Curtis, the *Climate Wayfinding* program has flourished because of your leadership, verve, and boundless heart. It is a privilege and total delight to work with you. Whit Bolster and Lydia Sweeney, your partnership in all things design is divine. Thank you for gracing this book with its vibe and beauty. To the leaders and poets who enliven these pages—thank you for your wisdom and for your generosity in sharing it.

Jennifer Herrera, I'm so grateful you could see the potential in an unusual book and for all you did, as my agent, to help it find its way. Melissa Zahorsky, thank you for giving me such a warm publishing home, as my editor, and rare creative freedom to shape this book *just so*. To everyone at Andrews McMeel—especially Julie Barnes, Riley Davis, Nicole Halper, Lynne McAdoo, Kirsty Melville, Tim Paulson, Kelsey Rolofson, Madison Schultz, Beth Steiner, Cat Vaughn, and Rima Weinberg—my gratitude for your talent and care. The bloom of this book includes seedling ideas planted many years ago; thank you, Naina Bajekal, David Biello, and Pat Mitchell, for the soil.

This book also has roots in a program and is stronger for all those who have helped that program grow. Laura Deaton, Alex Dileo, Carla Goldstein, Nicole Ng, Abby Schroering, and the whole Multiplier and Omega Institute teams—thank you for being comrades. Laura England, Kirsten Greenwald, Toby Herzlich, Sherri Mitchell, and Nikki Silvestri—thank you for brilliant feedback. Generous funding has made the *Climate Wayfinding* program possible. I'm especially grateful to those who have championed the work: Andrea and Karla Arria-Devoe (Mustang Foundation), Beth Braun (Libra Fund), Alex Conliffe (McCall MacBain Foundation), Holly Fogle and Caroline Noble (Monarch Foundation), Sara Hinkle (Towanda Foundation), Peter Tavernise and Alex Wilkins (Cisco Foundation), and Terri Taylor

(Lumina Foundation), who's been with us from the start. To all those who joined early cohorts, thank you for your trust, input, and encouragement. To a rich community of facilitators, thank you for breathing life into this program and sharing it with so many others.

Echoes of the wisdom and practice of many teachers are present in these pages. I feel especially thankful for learning from Deva Joy Gouss, Sherri Mitchell, Parker Palmer, Nikki Silvestri, and the Plum Village monastics, and for educators who shaped my path in profound ways: Susan Tinsley Daily and Ted Wesemann at the Outdoor Academy, Jane Pepperdene at the Paideia School, Gary Phillips and Jerry Smith at Sewanee, and Max Boykoff, Dave Frame, and Diana Liverman at Oxford. *Climate Wayfinding* is part of a rich ecosystem of kindred efforts, past and present.

Reflecting on my own climate journey, I feel such appreciation for those who have infused it with kindness, collaboration, and inspiration, in particular: Mary Bruce Alford, Emily Atkin, Lindsay Baker, Janine Benyus, Erin Bridges, John Cole, Kat Finlay, Jon Foley, Sam Gill, Heather Bird Harris, Mary Anne Hitt, Martha Jeffries, Kamal Kapadia, Kate Marvel, Jason Morris, Billy Parish, Mary Robinson, Jess Search, Megha Sood, Bernard Soubry, John Sutter, Brandon Sutton, Dar Vanderbeck, Bina Venkataraman, Dan Wilner, Britt Wray, and all the other dandelions. Leah Stokes, I feel damn lucky that a podcast brought us together. Ayana Elizabeth Johnson, thank you for what I can only call climate sistership.

To my wild and wonderful family, especially words-women Lucy Keeble and Janet Ward, thank you for believing in me. Jack Wilkinson, your departure broke my heart open, and in some aching way, this book is truer for it. (Perhaps slightly punnier too?) I love you, always. Ali Wilkinson, 🤟 my precious sissy. And to those who cared for me through the simultaneity of losing a parent and making a book, thank you: Romily Bernard, Alex Conliffe, Caroline Mauldin Dhane, Marissa Doran, Deva Joy Gouss, Laurel Graefe, Leah Naomi Green, Emma Lacey-Bordeaux, Jodi Mansbach, Michelle Maziar, Aditi Misra, Sherri Mitchell, Emily Crowe Pack, Jen Robinson, Rick Sauerman, and so many "Wilky" friends.

Lee, you are beyond. Thank you for being sun and refuge, for cheering me on and making sure I eat well. I am better in every way for sharing this life with you. Thank you, Arthur (dog) and Chi (cat), my cuddly and dedicated writing companions, and Birdie (horse), my sweetheart dance partner. Hawk, moon, oaks, and all the rest, you make the magic. And Earth—oh, Earth, *thank you.*

Guidelines for Gathering as a Group

These guidelines are vital for anyone shepherding a group through a collective *Climate Wayfinding* experience. If you're taking part in a group, you might find them useful too.

Bring your best to the space.

+ Model a warm, welcoming, open presence. How you show up begets how others will show up.
+ Take a moment to greet each person as they arrive. If there are people you're not well acquainted with, show interest and learn a bit about them. Small moments of connection can help everyone feel seen and at ease.
+ Practice active listening. This is one of the most powerful skills you can bring to a small group, as it encourages a shared spirit of attention and appreciation.
+ Breathe.

Cultivate a heartful space.

+ Strive to create a sense of belonging. However people are doing and whatever they are feeling, aim for everyone in the group to feel included and accepted just as they are.
+ Allow for pauses and silences. People often need a moment to reflect, gather their thoughts, and perhaps muster some courage. Intentional silence can deepen sharings.
+ Be prepared for expressions of emotion. Tears and laughter are both beautiful things. If someone seems overwhelmed by strong emotions, a gentle, empathetic question is often the best response: *How can I support you?* If what they need goes beyond what you can offer, guide them to additional resources at climatewayfinding.earth. ⊘

Hold a well-shared space.

+ Use the Group Agreements throughout the journey. As needed, you can invite people to reflect on these edges of your collective garden and address any that might need attention and tuning.
+ Encourage those who are quieter to join in. For example: *Let's hear from folks who haven't spoken yet.* Gentle eye contact, open body language, and pauses can encourage someone to participate, without putting them on the spot.
+ Guide those who are verbose to give ground. When introducing a whole-group element of the agenda (e.g., opening/closing circles), try this framing: *To allow time for all voices, please bring a spirit of brevity to your sharing.* If needed, you can refer back to it: *As a reminder, please aim for a spirit of brevity.*

Tune the experience.

+ Begin every session with music and poetry (read twice). You'll see this recurring rhythm in the agendas. It creates a threshold—a sense of entering into an experience that is distinct and special.
+ Role model how to respond. This is especially useful for opening and closing circles. If the prompt asks you to share one thing, share one thing. If it needs to be fairly short, keep it short. The group will take cues from you.
+ Play instrumental music during the *reflection* segments. Doing so can help create a relaxed and contemplative atmosphere. Find a curated selection of instrumental music at climatewayfinding.earth.
+ Be mindful of timing throughout. Without making the experience rigid, aim to wrap up each part of the agenda in the allotted time. That will ensure you can include all intended elements in the gathering and land it well at the end.

There is such sweetness in the act of bringing a group together. Look with wonder at how it grows you. Delight in what the group creates together. May the journey be rich with camaraderie and insight.

Essential Reading

These texts provided essential fiber for weaving this book. Find a digital version of this list with links at climatewayfinding.earth.

Baldwin, James. *Nobody Knows My Name: More Notes of a Native Son.* Dial Press, 1961.

Bass, Ellen. "The Big Picture." In *The Human Line.* Copper Canyon Press, 2007.

Berry, Wendell. *The Unforeseen Wilderness: An Essay on Kentucky's Red River Gorge.* University Press of Kentucky, 1971.

Beyer, Tamiko. "Equinox." In *Last Days.* Alice James Books, 2021.

brown, adrienne maree. *Holding Change: The Way of Emergent Strategy Facilitation and Mediation.* AK Press, 2021.

Cameron, Julia. *The Artist's Way: A Spiritual Path to Higher Creativity.* Jeremy P. Tarcher/Putnam, 1992.

Clifton, Lucille. "the mississippi river empties into the gulf." In *How to Carry Water: Selected Poems.* BOA Editions, 1996.

Harjo, Joy, and Stephen Strom. *Secrets from the Center of the World.* University of Arizona Press, 1989.

Hirshfield, Jane. "Optimism." In *Given Sugar, Given Salt: Poems.* HarperCollins, 2001.

hooks, bell. *All About Love: New Visions.* William Morrow, 2000.

Kimmerer, Robin Wall. *Braiding Sweetgrass: Indigenous Wisdom, Scientific Knowledge, and the Teachings of Plants.* Milkweed Editions, 2013.

Laméris, Danusha. "Nothing Wants to Suffer." In *Blade by Blade.* Copper Canyon Press, 2024.

Limón, Ada. "Dead Stars." In *The Carrying.* Milkweed Editions, 2018.

Lorde, Audre. "A Burst of Light: Living with Cancer." In *A Burst of Light and Other Essays*. Firebrand Books, 1988.

Marvel, Kate. *Human Nature: Nine Ways to Feel About Our Changing Planet*. Ecco, 2025.

Maathai, Wangari. "Nobel Lecture." Nobel Prize, December 10, 2004.

Meadows, Donella. "Leverage Points: Places to Intervene in a System." Sustainability Institute, 1999.

Milligan, Katherine, Juanita Zerda, and John Kania. "The Relational Work of Systems Change." *Stanford Social Innovation Review*, January 18, 2022.

Mitchell, Sherri. *Sacred Instructions: Indigenous Wisdom for Living Spirit-Based Change*. North Atlantic Books, 2018.

Morales, Aurora Levins. "Summons." In *RIMONIM: Ritual Poetry of Jewish Liberation*. Ayin Press, 2024.

O'Connor, M. R. *Wayfinding: The Science and Mystery of How Humans Navigate the World*. St. Martin's Press, 2019.

Palmer, Parker. *Let Your Life Speak: Listening for the Voice of Vocation*. Jossey-Bass, 2000.

Renkl, Margaret. *The Comfort of Crows: A Backyard Year*. Spiegel & Grau, 2023.

Sawin, Elizabeth. *Multisolving: Creating Systems Change in a Fractured World*. Island Press, 2024.

Smith, Tracy K. "An Old Story." In *Such Color: New and Selected Poems*. Graywolf Press, 2018.

Wray, Britt. *Generation Dread: Finding Purpose in an Age of Climate Crisis*. Knopf Canada, 2022.

Wild Chorus by Abacus Corvus was a mystically wise and profoundly generous companion as I created *Climate Wayfinding*.

Credits

Grateful acknowledgment to the following for permission to reprint previously published material:

Ada Limón, "Dead Stars," from *The Carrying*. Copyright © 2018 by Ada Limón. Reprinted with the permission of The Permissions Company, LLC on behalf of Milkweed Editions, milkweed.org.

"Climate Emotions Wheel" by Panu Pihkala, Anya Kamenetz, and Sarah Newman. Adapted and used by permission of the Climate Mental Health Network (Creative Commons BY-NC-SA 2024), climatementalhealth.net.

Climate Wayfinding program content. Copyright © 2024 by The All We Can Save Project. Adapted and used by permission of Multiplier on behalf of The All We Can Save Project, allwecansave.earth.

"Colette Pichon Battle—On Knowing What We're Called To," from *On Being with Krista Tippett*. Copyright © 2022 by Krista Tippett. Adapted and used by permission of The On Being Project, onbeing.org.

Danusha Laméris, "Nothing Wants to Suffer," from *Blade by Blade*. Copyright © 2024 by Danusha Laméris. Reprinted with the permission of The Permissions Company, LLC on behalf of Copper Canyon Press, coppercanyonpress.org.

Ellen Bass, "The Big Picture," from *The Human Line*. Copyright © 2007 by Ellen Bass. Reprinted with the permission of The Permissions Company, LLC on behalf of Copper Canyon Press, coppercanyonpress.org.

"Green New Career Types" © 2025 by the Sunrise Movement. Adapted and used by permission of the Sunrise Movement, sunrisemovement.org.

Lucille Clifton, "the mississippi empties into the gulf," from *How to Carry Water: Selected Poems*. Copyright © 1996 by Lucille Clifton. Reprinted with the permission of The Permissions Company, LLC on behalf of BOA Editions Ltd., boaeditions.org.

"Optimism" from *Given Sugar, Given Salt* by Jane Hirshfield. Copyright © 2001 by Jane Hirshfield. Used by permission of HarperCollins Publishers.

"Summons" from *RIMONIM: Ritual Poetry of Jewish Liberation* by Aurora Levins Morales (Ayin Press, 2024). Copyright © 2024 by Aurora Levins Morales. Used by permission of Anderson Literary Management LLC. All rights reserved.

Tamiko Beyer, "Equinox," from *Last Days*. Copyright © 2021 by Tamiko Beyer. Reprinted with the permission of The Permissions Company, LLC on behalf of Alice James Books, alicejamesbooks.org.

Tracy K. Smith, "An Old Story," from *Such Color: New and Selected Poems*. Copyright © 2018 by Tracy K. Smith. Reprinted with the permission of The Permissions Company, LLC on behalf of Graywolf Press, Minneapolis, Minnesota, graywolfpress.org.

The leaders featured in "Lighting the Way" stories generously shared their wisdom and permission to be included in this book. Thank you to Colette Pichon Battle, Valérie Courtois, Wahleah Johns, Kate Marvel, Wanjira Mathai, Sherri Mitchell, Billy Parish, Jennifer Robinson, and Sister True Dedication.

This book also draws upon these previously published works from the author:

All We Can Save: Truth, Courage, and Solutions for the Climate Crisis. One World, 2020. (co-editor with Ayana Elizabeth Johnson)

Between God & Green: How Evangelicals Are Cultivating a Middle Ground on Climate Change. Oxford University Press, 2012.

"The Climate Crisis Is a Call to Action. These 5 Steps Helped Me Figure Out How to Be of Use." *Time*, July 19, 2021.

Drawdown: The Most Comprehensive Plan Ever Proposed to Reverse Global Warming. Penguin Books, 2017. (lead writer)

The Drawdown Review: Climate Solutions for a New Decade. Project Drawdown, 2020. (editor and writer)

"How Empowering Women and Girls Can Help Stop Global Warming." TED Talk, Palm Springs, CA, November 2018.

A Matter of Degrees. Select podcast episodes, 2020–2023. (co-host with Leah Stokes)

"Opinion: What Threatens My 'City in the Forest.'" *CNN*, August 28, 2023.

"Pinched Between Peril and Possibility, We Need Climate Leadership." *Atlanta Journal-Constitution*, October 17, 2024.

"7 Resources to Help You Cope with Climate Anxiety." *Time*, November 3, 2021. (co-author with Britt Wray)

"Women, Girls, and Non-Binary Leaders Are Demonstrating the Kind of Leadership Our World So Badly Needs." *The Elders*, December 6, 2019.

Notes

Find a digital version of these notes with links at climatewayfinding .earth.

"And the world cannot be discovered": Wendell Berry, *The Unforeseen Wilderness: An Essay on Kentucky's Red River Gorge* (University Press of Kentucky, 1971), 34.

Pre-Ambling

"Come closer. / Come into this": Anis Mojgani, "Closer," in *Songs from Under the River* (Write Bloody Publishing, 2013), 21.

1. Embarking

Qualla Boundary, the 56,600-acre territory: "About Us," Visit Cherokee, NC, Eastern Band of Cherokee Indians.

"morning pages," one of Julia Cameron's two essential practices: Julia Cameron, *The Artist's Way: A Spiritual Path to Higher Creativity* (Jeremy P. Tarcher/Putnam, 1992).

"Tell me, what is it you plan to do": Mary Oliver, "The Summer Day," in *New and Selected Poems, Volume One* (Beacon Press, 1992), 94.

warmest temperatures in at least one hundred thousand years: Darrell S. Kaufman and Nicholas P. McKay, "Technical Note: Past and Future Warming—Direct Comparison on Multi-Century Timescales," *Climate of the Past* 18 (2022): 911–17.

Globally, 84 percent of young people: Caroline Hickman et al., "Climate Anxiety in Children and Young People and Their Beliefs About Government Responses to Climate Change: A Global Survey," *Lancet Planetary Health* 5, no. 12 (December 2021): 863–73.

A majority of Americans are worried, too: Anthony Leiserowitz et al., "Climate Change in the American Mind: Beliefs & Attitudes, Spring 2025," Yale Program on Climate Change Communication, 2025.

a quarter of the country expresses alarm: Anthony Leiserowitz et al., "Global Warming's Six Americas, Fall 2024," Yale Program on Climate Change Communication, 2025.

a much smaller group—just 8 percent of us: Matthew Goldberg et al., "Segmenting the Climate Change Alarmed: Active, Willing, and Inactive," Yale Program on Climate Change Communication, 2021.

Around the world, 89 percent of people: Peter Andre et al., "Globally Representative Evidence on the Actual and Perceived Support for Climate Action," *Nature Climate Change* 14 (2024): 253–59.

"We are living in high-stakes times": Makani Themba, "Stepping Up, Stepping In: Facilitating for Freedom," in *Holding Change: The Way of Emergent Strategy Facilitation and Mediation,* ed. adrienne maree brown (AK Press, 2021), 56.

In the 1800s, scientists like Eunice Newton Foote: "History of Climate Science Research," UCAR Center for Science Education.

"Why are the keys to our future in the hands": Andrea Gibson, "Homesick: a Plea for Our Planet," in *You Better Be Lightning* (Button Poetry, 2021).

"the climate crisis is a leadership crisis": Suyin Haynes, "Meet 15 Women Leading the Fight Against Climate Change: Katharine Wilkinson: Education," *Time,* September 12, 2019; Katharine Wilkinson, "Women, Girls, and Non-Binary Leaders Are Demonstrating the Kind of Leadership Our World So Badly Needs," *The Elders,* December 6, 2019; Ayana Elizabeth Johnson and Katharine K. Wilkinson, "Begin," in *All We Can Save: Truth, Courage, and Solutions for the Climate Crisis* (One World, 2020), xvii–xxiv.

Just 37 percent of American adults: Gregory A. Smith et al., "Religious Attendance and Congregational Involvement," Pew Research Center, February 26, 2025.

the true complexity of that request: Katharine Wilkinson, "How Empowering Women and Girls Can Help Stop Global Warming," TED Talk, Palm Springs, CA, November 2018.

a five-step approach to answering the question of *What can I do?*: Katharine Wilkinson, "The Climate Crisis Is a Call to Action. These 5 Steps Helped Me Figure Out How to Be of Use," *Time,* July 19, 2021.

"the act of finding one's way to a particular place": *Oxford English Dictionary,* "wayfinding."

In Indigenous wayfinding traditions . . . signs and signals guide navigation: M. R. O'Connor, *Wayfinding: The Science and Mystery of How Humans Navigate the World* (St. Martin's Press, 2019).

the term *more-than-human world*: David Abram, *The Spell of the Sensuous: Perception and Language in a More-Than-Human World* (Pantheon Books, 1996).

metaphor of the air held between a drum's round shell and taut skin: Sherri Mitchell, speech, Global Matriarchs' Gathering, Wicuhkemtultine Kinship Community, Maine, September 24, 2022.

Colette Pichon Battle's story, adapted with permission from: Colette Pichon Battle, "On Knowing What We're Called To," interview by Krista Tippett, *On Being with Krista Tippett,* podcast, The On Being Project, March 3, 2022.

2. Getting Our Bearings

roughly 50 gigatons of greenhouse pollution: Alfredo Rivera, Shweta Movalia, and Emma Rutkowski, "Global Greenhouse Gas Emissions: 1990–2022 and Preliminary 2023 Estimates," Rhodium Group, November 26, 2024.

Judith Viorst's classic tale: Judith Viorst, *Alexander and the Terrible, Horrible, No Good, Very Bad Day* (Atheneum, 1972).

burgeoning community of climate leaders within American evangelical Christianity: Katharine K. Wilkinson, *Between God & Green: How Evangelicals Are Cultivating a Middle Ground on Climate Change* (Oxford University Press, 2012).

The Systems Change Lab defines *systems change*: "What Is Systems Change?," Systems Change Lab.

1.4 degrees Celsius: C3S Global Temperature Trend Monitor, Copernicus Climate Change Service, October 2025.

Kate Marvel's story, adapted with her collaboration and permission from: Kate Marvel, written exchange with the author, July 28, 2025; Kate Marvel, *Human Nature: Nine Ways to Feel About Our Changing Planet* (Ecco, 2025); Kate Marvel, "What Have We Learned from a Summer of Climate Reckoning?," interview by David Wallace-Wells, *The Ezra Klein Show, New York Times,* September 5, 2023; Kate Marvel, "The End of Normal: Understanding—and Correcting—Earth's Troubling Climate Trajectory," webinar, posted August 28, 2023, by Project Drawdown, YouTube.

"the rarest and purest form of generosity": Simone Weil to Joë Bousquet, April 13, 1942, in Simone Pétrement, *Simone Weil: A Life,* trans. Raymond Rosenthal (Pantheon Books, 1976), 462.

3. Following Our Emotions

Google searches for "climate anxiety" soared 565 percent: Kate Yoder, "It's Not Just You: Everyone Is Googling 'Climate Anxiety,'" *Grist,* October 28, 2021.

nearly half of young people surveyed: Caroline Hickman et al., "Climate Anxiety in Children and Young People and Their Beliefs About Government Responses to Climate Change: A Global Survey," *Lancet Planetary Health* 5, no. 12 (December 2021): 863–73.

What Dr. Lise Van Susteren, a practicing psychiatrist, termed *pre-traumatic stress*: Lise Van Susteren, "Our Children Face 'Pretraumatic Stress' from Worries About Climate Change," *BMJ Opinion,* November 19, 2020.

more than eight billion human beings: United Nations, Department of Economic and Social Affairs, Population Division, *World Population Prospects 2024: Summary of Results,* DESA/POP/2024/TR/NO. 9 (2024).

over two million known species: O. Bánki et al., *Catalogue of Life*, Version 2025-10-10 XR, 2025.

climate anxiety . . . healthy response to an existential threat: Britt Wray, *Generation Dread: Finding Purpose in an Age of Climate Crisis* (Knopf Canada, 2022).

"taxonomy of climate emotions": Panu Pihkala, "Toward a Taxonomy of Climate Emotions," *Frontiers in Climate* 3 (2022).

Climate Emotions Wheel: Panu Pihkala, Anya Kamenetz, and Sarah Newman, "Climate Emotions Wheel," Climate Mental Health Network, 2024.

"The planet, like us, has lost much before": Kate Marvel, *Human Nature: Nine Ways to Feel About Our Changing Planet* (Ecco, 2025), 120.

"There must be those among whom we can sit down and weep": Adrienne Rich, "Sources," in *Your Native Land, Your Life: Poems* (W. W. Norton, 1986), 25.

"fucking face our grief" . . . "shallow actions": Kritee Kranko, "How to Cope with All the Climate Feels," interview by Katharine Wilkinson, *A Matter of Degrees*, podcast, season 3, episode 4, November 2, 2022.

"Our feelings are our most genuine paths": Audre Lorde, interview by Claudia Tate, in *Black Women Writers at Work,* ed. Claudia Tate (Oldcastle Books, 1985), 106–7.

not a sign of something wrong with us: Sherri Mitchell, "How Pain Can Drive Climate Action," posted November 27, 2020, by Omega Institute for Holistic Studies, YouTube.

Participating in collective action: Sarah E. O. Schwartz et al., "Climate Change Anxiety and Mental Health: Environmental Activism as Buffer," *Current Psychology* 42 (2023): 16708–21.

She began to edit Thich Nhat Hanh's books on Buddhism and ecology: Thich Nhat Hanh, *Love Letter to the Earth* (Parallax Press, 2012); Thich Nhat Hanh, *Zen and the Art of Saving the Planet* (HarperOne, 2022).

Sister True Dedication's story, adapted with her collaboration and permission from: Sister True Dedication, in discussion with the author, July 2, 2025; Sister True Dedication, "3 Questions to Build Resilience—and Change the World," TED Talk, Edinburgh, Scotland, October 2021; Sister True Dedication, guided meditation from "Zen and the Art of Saving the Planet, Week 7: Action Dimension," online course, Plum Village, 2023.

"sponge of gratitude": Ross Gay, "Catalog of Unabashed Gratitude," in *Catalog of Unabashed Gratitude* (University of Pittsburgh Press, 2015), 93.

4. Surfacing Solutions

"I don't believe any longer" . . . "we have to make it over": James Baldwin, address at the *Esquire* symposium, San Francisco State College, October 22, 1960, Poetry Center Digital Archive, audio and transcript; James Baldwin, "Notes for a Hypothetical Novel," in *Nobody Knows My Name: More Notes of a Native Son* (Dial Press, 1961), 126.

major fossil fuel companies . . . conditions of the last ten thousand years: Beth Gardiner, "How an Early Oil Industry Study Became Key in Climate Lawsuits," *Yale Environment 360,* November 30, 2022; Oliver Milman, "'Smoking Gun Proof': Fossil Fuel Industry Knew of Climate Danger as Early as 1954, Documents Show," *The Guardian* (US edition), January 30, 2024.

I spent most of 2016 at my desk, writing most of the book *Drawdown*: Paul Hawken, ed., *Drawdown: The Most Comprehensive Plan Ever Proposed to Reverse Global Warming* (Penguin Books, 2017).

"One of the best and most challenging things": Rebecca Solnit, "What Can I Do About the Climate Emergency?," digital addendum to *Not Too Late: Changing the Climate Story from Despair to Possibility,* eds. Rebecca Solnit and Thelma Young Lutunatabua (Haymarket Books, 2023).

the follow-up publication I led, *The Drawdown Review*: Katharine Wilkinson, ed., *The Drawdown Review: Climate Solutions for a New Decade* (Project Drawdown, 2020).

roughly the combined weight of all livestock (mostly cattle) and all human beings on Earth: Lior Greenspoon et al., "The Global Biomass of Wild Mammals," *Proceedings of the National Academy of Sciences* 120, no. 10 (2023): e2204892120.

"Leverage points are points of power": Donella Meadows, "Leverage Points: Places to Intervene in a System," Sustainability Institute, 1999.

I saw that in action . . . efforts to build a regenerative economy: Leah Stokes and Katharine Wilkinson, hosts, *A Matter of Degrees*, podcast, season 3, episode 10, "The Tongass: A Way Forward for the Forest," March 2, 2023.

the Tongass is the largest intact temperate rainforest in the world: "The Tongass National Forest: America's Climate and Salmon Forest," Sitka Conservation Society, October 5, 2022.

"We brought down apartheid": "Our Two Priorities: The Climate Emergency and the Crisis in Democracy," Doc Society.

a "wild rumpus," reminiscent of Maurice Sendak's fantastical scenes: Maurice Sendak, *Where the Wild Things Are* (Harper & Row, 1963).

Or, for a more recent reference: *Everything Everywhere All at Once,* written and directed by Daniel Kwan and Daniel Scheinert, A24, 2022.

Andrew Boyd proposes three questions for solution evaluation: Andrew Boyd, *I Want a Better Catastrophe: Navigating the Climate Crisis with Grief, Hope, and Gallows Humor* (New Society Publishers, 2023), 231–32.

Dr. Elizabeth "Beth" Sawin calls this *multisolving*: Elizabeth Sawin, *Multisolving: Creating Systems Change in a Fractured World* (Island Press, 2024).

"The bottom line is this": John Romano, "James Baldwin Writing and Talking," *New York Times*, September 23, 1979.

"a more human dwelling place": James Baldwin, "The Creative Process," in *Creative America*, ed. Jerry Mason (Ridge Press, 1962), 17.

"The majority of the world's remaining biodiversity is on lands cared for and loved by Indigenous peoples": Katie Reytar, Peter Veit, and Johanna von Braun, "Protecting Biodiversity Hinges on Securing Indigenous and Community Land Rights," *Insights*, World Resources Institute, November 22, 2024; Latoya Abulu, Aimee Gabay, and Sonam Lama Hyolmo, "Do Indigenous Peoples Really Conserve 80% of the World's Biodiversity?," *Mongabay*, September 26, 2024.

boreal forest that covers more than half of Canada: *The Canadian Encyclopedia*, "Boreal Zone," last updated May 25, 2018.

It is the largest intact forest ecosystem on Earth: Michelle Sims, Peter Potapov, and Elizabeth Goldman, "The World's Last Intact Forests Are Becoming Increasingly Fragmented," *Insights*, World Resources Institute, November 2, 2022.

Valérie Courtois's story, adapted with her collaboration and permission from: Valérie Courtois, in discussion with the author, July 3, 2025; Valérie Courtois, "How Indigenous Guardians Protect the Planet and Humanity," TED Talk, New York, NY, September 2022; Valérie Courtois, "Bright Award Event and Discussion Honoring 2023 Winner Valérie Courtois," interview by Shawn Watts, October 2, 2023, posted October 5, 2023, by Stanford Law School, YouTube; David McGuffin, "Valérie Courtois on What She Hopes Will Come out of COP15: 'To Save the World,'" *Canadian Geographic*, December 9, 2022.

5. Offering Our Superpowers

As I wrote for *CNN*, being the "city in a forest": Katharine K. Wilkinson, "Opinion: What Threatens My 'City in the Forest,'" *CNN*, August 28, 2023.

"Every living thing": Margaret Renkl, *The Comfort of Crows: A Backyard Year* (Spiegel & Grau, 2023), 78–79.

"No matter how sick I feel": Audre Lorde, "A Burst of Light: Living with Cancer," in *A Burst of Light and Other Essays* (Firebrand Books, 1988), 76–77.

"Even a wounded world is feeding us": Robin Wall Kimmerer, *Braiding Sweetgrass: Indigenous Wisdom, Scientific Knowledge, and the Teachings of Plants* (Milkweed Editions, 2013), 327.

"Perhaps this / is [life's] way of fighting back": Mary Oliver, "Don't Hesitate," in *Devotions: The Selected Poems of Mary Oliver* (Penguin Press, 2017), 61.

"every job is a climate job": Jamie Alexander, "No Matter Where We Work, Every Job Is a Climate Job Now," TED Talk, TEDxClimateActionTech, December 2020.

American Climate Corps, a national service program maddeningly axed in early 2025: Kate Yoder, "The American Climate Corps Is Over. What Even Was It?," *Grist*, January 15, 2025.

what Joanna Macy called the *Great Turning*: Joanna Macy and Chris Johnstone, *Active Hope: How to Face the Mess We're in Without Going Crazy* (New World Library, 2012); Joanna Macy and Jessica Serrante, hosts, *We Are the Great Turning*, podcast, 2024, Sounds True.

Peabody pulled more than a billion gallons of pristine groundwater: Tim Grabiel, "Drawdown: An Update on Groundwater Mining on Black Mesa," *NRDC Issue Paper*, Natural Resources Defense Council, March 2006.

Wahleah Johns and Billy Parish's story, adapted with their collaboration and permission from: Wahleah Johns and Billy Parish, in discussion with the author, August 13, 2025; Billy Parish and Dev Aujla, *Making Good: Finding Meaning, Money, and Community in a Changing World* (Rodale Books, 2012), 12–16; Wahleah Johns, "Changing Woman: One Navajo's Fight for a Just Energy Transition," interview by Julian Brave NoiseCat, *A Matter of Degrees*, podcast, season 1, episode 7, December 8, 2020; Billy Parish, "742: Mosaic's Journey to $15 Billion in Residential Solar Loans with Co-Founder & Executive Chair, Billy Parish," interview by Nico Johnson, *SunCast*, episode 742, Spotify, September 19, 2024; Wahleah Johns, "Wahleah Johns and the Movement to Replace Extraction with Regeneration," interview by May Boeve, *How We Build This*, podcast, episode 4, Spotify, July 15, 2025.

6. Connecting with Community

It forms more compounds than all the other elements: *Britannica*, "carbon."

"I contain multitudes": Walt Whitman, "Song of Myself," in *Leaves of Grass* (David McKay, 1892), 78.

"Everything we know about systems tells us relationships are the core": Katherine Milligan, Juanita Zerda, and John Kania, "The Relational Work of Systems Change," *Stanford Social Innovation Review*, January 18, 2022.

One definition of *life*: David Haskell, "How Listening to Trees Can Help Reveal Nature's Connections," interview by Diane Toomey, *Yale Environment 360*, August 4, 2017; David Haskell, written exchange with the author, August 30, 2025.

"conscious emulation of nature's genius": Janine M. Benyus, *Biomimicry: Innovation Inspired by Nature* (HarperCollins, 1997), 2.

"Be locally attuned and responsive": "DesignLens: Life's Principles," Biomimicry 3.8.

"The most important thing an individual can do": Bill McKibben, "Give Up Your Climate Guilt," interview by Leah Stokes, *A Matter of Degrees*, podcast, season 1, episode 1, October 14, 2020.

Dr. Parker Palmer's small, wise book: Parker Palmer, *Let Your Life Speak: Listening for the Voice of Vocation* (Jossey-Bass, 2000).

"Rarely, if ever, are any of us healed in isolation": bell hooks, *All About Love: New Visions* (William Morrow, 2018), 215.

Parker's concept of a *community of congruence*: Parker J. Palmer, *The Courage to Teach: Exploring the Inner Landscape of a Teacher's Life* (Jossey-Bass, 1998), 173–84.

"get in good trouble, necessary trouble": John Lewis, afterword to *His Truth Is Marching On: John Lewis and the Power of Hope,* by Jon Meacham (Random House, 2020), 248.

the *All We Can Save* anthology: Ayana Elizabeth Johnson and Katharine K. Wilkinson, eds., *All We Can Save: Truth, Courage, and Solutions for the Climate Crisis* (One World, 2020).

"Go . . . where your heart can find a home": Ayana Elizabeth Johnson, *What If We Get It Right? Visions of Climate Futures* (One World, 2024), 427.

This is the country that has produced the most climate pollution over time: Matthew W. Jones et al., "National Contributions to Climate Change Due to Historical Emissions of Carbon Dioxide, Methane, and Nitrous Oxide Since 1850," *Scientific Data* 10 (2023): 155.

Kimberlé W. Crenshaw coined the term *intersectionality* in 1989: Kimberlé W. Crenshaw, "Demarginalizing the Intersection of Race and Sex: A Black Feminist Critique of Antidiscrimination Doctrine, Feminist Theory, and Antiracist Politics," *University of Chicago Legal Forum* 1989, no. 1 (1989): 139–67.

"values that reflect life": Tara Houska, "Sacred Resistance," in Johnson and Wilkinson, *All We Can Save,* 218.

fundamental causes of the sickness we are experiencing: Nadine Anne Hura, "Those Riding Shotgun," *PEN Transmissions,* May 6, 2021.

"There is an art to flocking": adrienne maree brown, *Emergent Strategy: Shaping Change, Changing Worlds* (AK Press, 2017), 13.

Up to about sixty thousand years, carbon can tell us: Steve Koppes and Louise Lerner, "Carbon-14 Dating, Explained," *UChicago News,* February 2025.

That absence makes it possible to identify their gaseous remains: Heather Graven, Ralph F. Keeling, and Joeri Rogelj, "Changes to Carbon Isotopes in Atmospheric CO_2 over the Industrial Era and into the Future," *Global Biogeochemical Cycles* 34, no. 11 (2020).

"We are called to assist the Earth": Wangari Maathai, "Nobel Lecture," Nobel Prize, December 10, 2004, video and transcript.

Of the fifty most climate-vulnerable countries in the world: Notre Dame Global Adaptation Initiative (ND-GAIN) Country Index, 2025.

Wanjira Mathai's story, adapted with her collaboration and permission from: Wanjira Mathai, written exchange with the author, August 1, 2025; Wanjira Mathai, "The Tree-Growing Movement Restoring Africa's Vital Landscapes," TED Talk, Vancouver, BC, April 2023; Wanjira Mathai, "How Gender Equality Can Save the Planet," interview by Ayana Elizabeth Johnson and Katharine Wilkinson, *A Matter of Degrees,* podcast, season 2, episode 10, September 29, 2021; Wanjira Mathai, "Wanjira Mathai—a New Story for Africa," interview by Amy Davoren-Rose and Angus Hervey, *Hope Is a Verb,* podcast, Spotify, July 21, 2023; Wanjira Mathai, "The Green Belt with Wanjira Mathai," interview by Christiana Figueres, Tom Rivett-Carnac, and Paul Dickinson, *Outrage + Optimism,* podcast, season 1, episode 33, December 5, 2019.

7. Orienting with Vision

"The task of stories is to teach us to see": Margaret W. Pepperdene, "The Figure of the Journey," lecture, September 1, 1999, Paideia School, author's notes.

The hippocampus, so named for its seahorse shape: Shyamal C. Bir et al., "Julius Caesar Arantius (Giulio Cesare Aranzi, 1530–1589) and the Hippocampus of the Human Brain: History Behind the Discovery," *Journal of Neurosurgery* 122, no. 4 (2015): 971–75.

It's famously large for long-tenured London taxi drivers: Eleanor A. Maguire et al., "Navigation-Related Structural Change in the Hippocampi of Taxi Drivers," *Proceedings of the National Academy of Sciences* 97, no. 8 (2000): 4, 398–403.

"The human imagination is like a beacon": M. R. O'Connor, *Wayfinding: The Science and Mystery of How Humans Navigate the World* (St. Martin's Press, 2019), 274.

Nikki Silvestri teaches that vision sources from our spirit or soul: Nikki Silvestri, lecture from "Joy + Impact Academy," online course, Soil and Shadow, 2021, author's notes.

The first compass was invented in ancient China, during the Han dynasty: Susan Silverman, "Compass, China, 220 BCE," Museum of Ancient Inventions, Smith College Program in the History of the Sciences.

the opening ceremony of the Beijing Summer Olympics celebrated it: "Incredible Highlights—Beijing 2008 Olympics, Opening Ceremony," posted January 19, 2010, by Olympics, YouTube.

This earliest compass . . . a means of divination: Silverman, "Compass, China, 220 BCE."

the compass was adapted for terrestrial and maritime navigation: "Technological Advances During the Song," China in 1000 CE: The Most Advanced Society in the World, Asia for Educators, Columbia University.

the poles of our planet . . . sit some 7,900 miles apart: John W. Foster, "Earth's Shape," EBSCO Research Starters, 2025.

That magnetic field . . . shelters us from the bombardment of solar wind and storms: Alan Buis, "Earth's Magnetosphere: Protecting Our Planet from Harmful Space Energy," NASA, August 3, 2021.

"compass of the heart": Tara Brach, host, *Tara Brach,* podcast, "Compass of the Heart," Vimeo, June 29, 2011.

"In order to find our way": Bayo Akomolafe, "Bayo Akomolafe: To Find Your Purpose, Become Lost," posted March 23, 2023, by Purpose Guides Institute, YouTube.

"Sometimes being lost, if there is such a thing, is the sweetest place to be": Linda Hogan, *The Woman Who Watches over the World: A Native Memoir* (W. W. Norton, 2001), 15.

navigation is primordial: O'Connor, *Wayfinding,* 274.

Fracking, or hydraulic fracturing: Melissa Denchak, "Fracking 101," Natural Resources Defense Council, April 19, 2019.

"to bring the world's biggest problem to the world's highest court": Julian Aguon, "The Fight for Climate Justice at the World Court," *Rolling Stone,* June 27, 2025.

"Land is our mother": Cynthia Houniuhi, "Cynthia's Oral Statement to the International Court of Justice," December 2, 2024, posted January 14, 2025, by Pacific Islands Students Fighting Climate Change, YouTube.

"The judges heard ninety-six government submissions": Jennifer Bansard et al., "Summary of the International Court of Justice Hearings on States' Obligations in Respect of Climate Change: 2–13 December 2024," *Earth Negotiations Bulletin,* International Institute for Sustainable Development, December 16, 2024.

"In the Pacific, we have always looked to the stars": Vishal Prasad, "Vishal's Oral Statement to the International Court of Justice," December 13, 2024, posted January 14, 2025, by Pacific Islands Students Fighting Climate Change, YouTube.

"has an important but ultimately limited role": "Obligations of States in Respect of Climate Change," International Court of Justice, Advisory Opinion (July 23, 2025), ¶ 456.

Jennifer Robinson's story, adapted with her collaboration and permission from: Jennifer Robinson, in discussion with the author, July 30, 2025; Jennifer Robinson, "Living Greatly in the Law: Law and Progressive Social Change," 2021 Hal Wootten Lecture, October 13, 2021, UNSW Sydney, audio and transcript; "CityTalks: The Climate and Nature Crisis—What Australia Does Matters!," April 2, 2025, posted April 9, 2025, by City of Sydney, YouTube.

"Release perfection": adrienne maree brown, *Holding Change: The Way of Emergent Strategy Facilitation and Mediation* (AK Press, 2021), 129.

8. Journeying Onward

"During the time that I'm here": Christiana Figueres, "Christiana Figueres: Ecological Hope, and Spiritual Evolution," interview by Krista Tippett, *On Being with Krista Tippett,* podcast, The On Being Project, November 9, 2023.

Flamenco was born . . . persecuted by Spain's Catholic elite: Donn E. Pohren, *The Art of Flamenco* (Bold Strummer, 2005); Giles Tremlett, *Ghosts of Spain: Travels Through Spain and Its Silent Past* (Walker Books, 2008).

"Gutsy" and "earthy": *¡Colores!,* season 19, episode 10, "Flamenco Dancer María Benítez," aired March 28, 2013, on New Mexico PBS.

Alaska issues its first-ever heat advisories: Lois Parshley, "Alaska Just Hit a Climate Milestone—Its First-Ever Heat Advisory," *Grist,* June 16, 2025.

the Klamath River, undammed, runs free from Oregon to California: John Branch, "First Time in 100 Years: Young Kayakers on a Ride for the Ages," *New York Times,* June 17, 2025.

Gas turbines powering the world's largest supercomputer foul the air of southwest Memphis, Tennessee: "We Went to the Town Elon Musk Is Poisoning," posted May 30, 2025, by More Perfect Union, YouTube.

a conservation group buys, and blocks, a mining site that threatened to imperil Georgia's Okefenokee Swamp: Kala Hunter, "Okefenokee Mining Site Purchased for $60M. Land Buyers Credit Public Advocacy," *Columbus Ledger-Enquirer,* June 24, 2025.

In it, we explore the personal, professional, and political dimensions: Leah Stokes and Katharine Wilkinson, hosts, *A Matter of Degrees,* podcast, season 3, episodes 1–3, 2022.

They are nine critical processes . . . "safe operating space for humanity": Johan Rockström et al., "A Safe Operating Space for Humanity," *Nature* 461 (2009): 472–75.

seven of nine boundaries are outstripped: N. H. Kitzmann et al., eds., *Planetary Health Check 2025: A Scientific Assessment of the State of the Planet* (Potsdam Institute for Climate Impact Research, 2025).

This particular rift has been opening: Alyssa L. Abbey and Nathan A. Niemi, "Perspectives on Continental Rifting Processes from Spatiotemporal Patterns of Faulting and Magmatism in the Rio Grande Rift, USA," *Tectonics* 39, no. 1 (2019).

Equisetum . . . is the only surviving genus of a primordial lineage: "Field Horsetail," CU Museum of Natural History, University of Colorado Boulder, August 12, 2020; *Britannica,* "horsetail."

"On the last day of the world": W. S. Merwin, "Place," in *The Rain in the Trees* (Knopf, 1988), 64.

Sherri Mitchell's story, adapted with her collaboration and permission from: Sherri Mitchell, in discussion with the author, June 25, 2025; Sherri Mitchell, *Sacred Instructions: Indigenous Wisdom for Living Spirit-Based Change* (North Atlantic Books, 2018), 3–5.

"I have lost my way many times": Joy Harjo and Stephen Strom, *Secrets from the Center of the World* (University of Arizona Press, 1989), 52.

Index

Note: Page numbers in *italics* indicate illustrated frameworks. Page numbers paired with an *n* indicate footnotes.

D

E

M

N

O

P

Q

R

S

Beyond these pages, find digital content
to enrich the journey at climatewayfinding.earth.

Enjoy *Climate Wayfinding* as an audiobook narrated
by the author, wherever audiobooks are sold.

About the Author

Dr. Katharine K. Wilkinson is a human on Earth. As a writer, teacher, and creator, she has inspired hundreds of thousands of climate journeys through transformational projects that shift our cultural narratives about what's possible and nurture engagement in renewing our world. Her publications include the bestselling anthology *All We Can Save*, the podcast *A Matter of Degrees*, and the *New York Times* bestseller *Drawdown*. Dr. Wilkinson co-founded and leads The All We Can Save Project, where she shaped the much-beloved programs All We Can Save Circles and Climate Wayfinding. She holds a DPhil in geography and environment from the University of Oxford, where she was a Rhodes Scholar, and a BA in religion from Sewanee: The University of the South. In 2019, *Time* magazine named her one of fifteen "Women Who Will Save the World." She lives with her loves in Atlanta, Georgia, and finds her deepest joy on a mountain or a horse.

All We Can Save: Truth, Courage, and Solutions for the Climate Crisis (co-editor with Ayana Elizabeth Johnson)

Between God & Green: How Evangelicals Are Cultivating a Middle Ground on Climate Change (author)

Drawdown: The Most Comprehensive Plan Ever Proposed to Reverse Global Warming (lead writer)

The Drawdown Review: Climate Solutions for a New Decade (editor and writer)

Human on Earth (Substack newsletter, author)

A Matter of Degrees (podcast, co-host with Leah Stokes)

Learn more at kkwilkinson.com.

Photo: Gabriella Valladeres

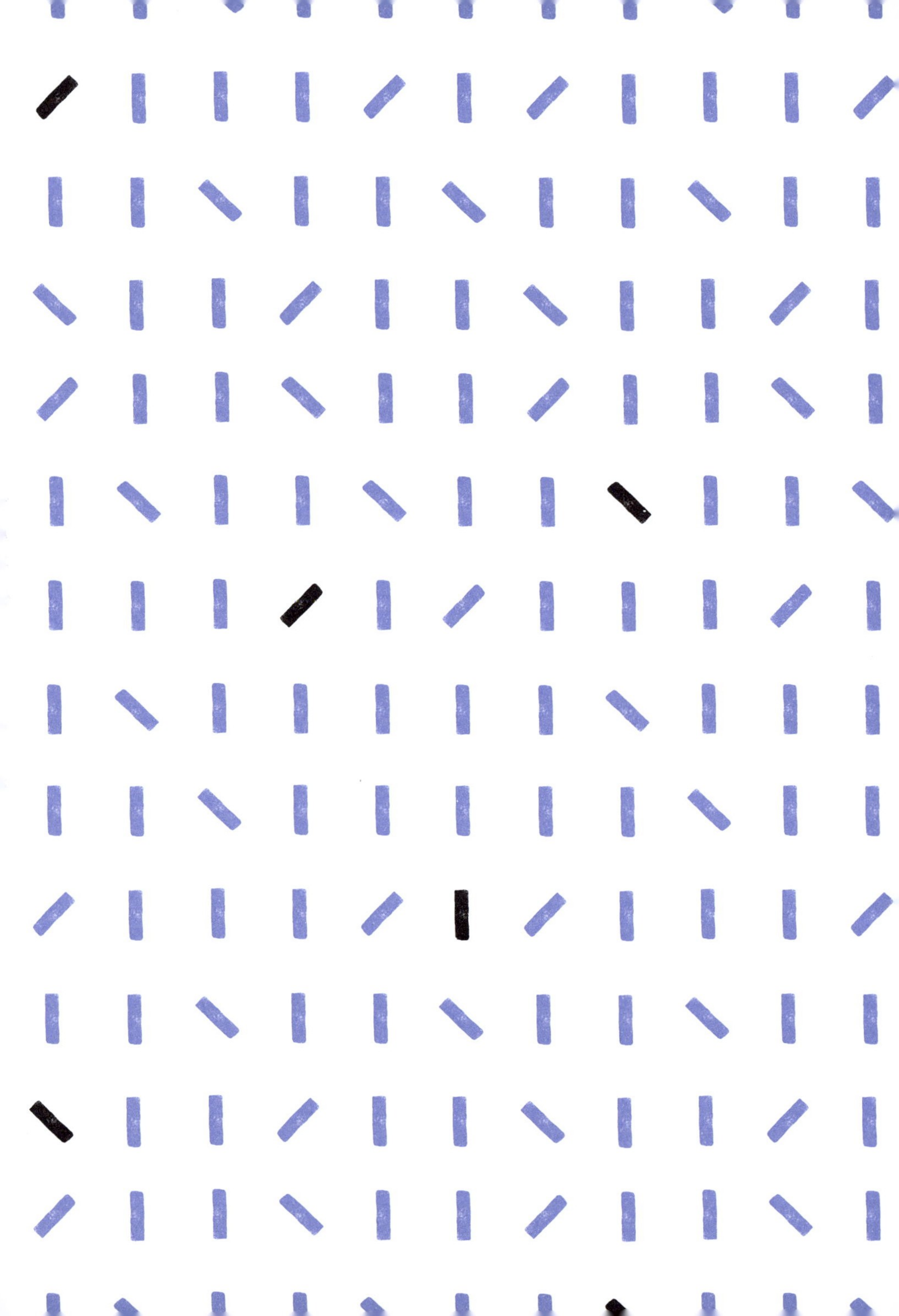

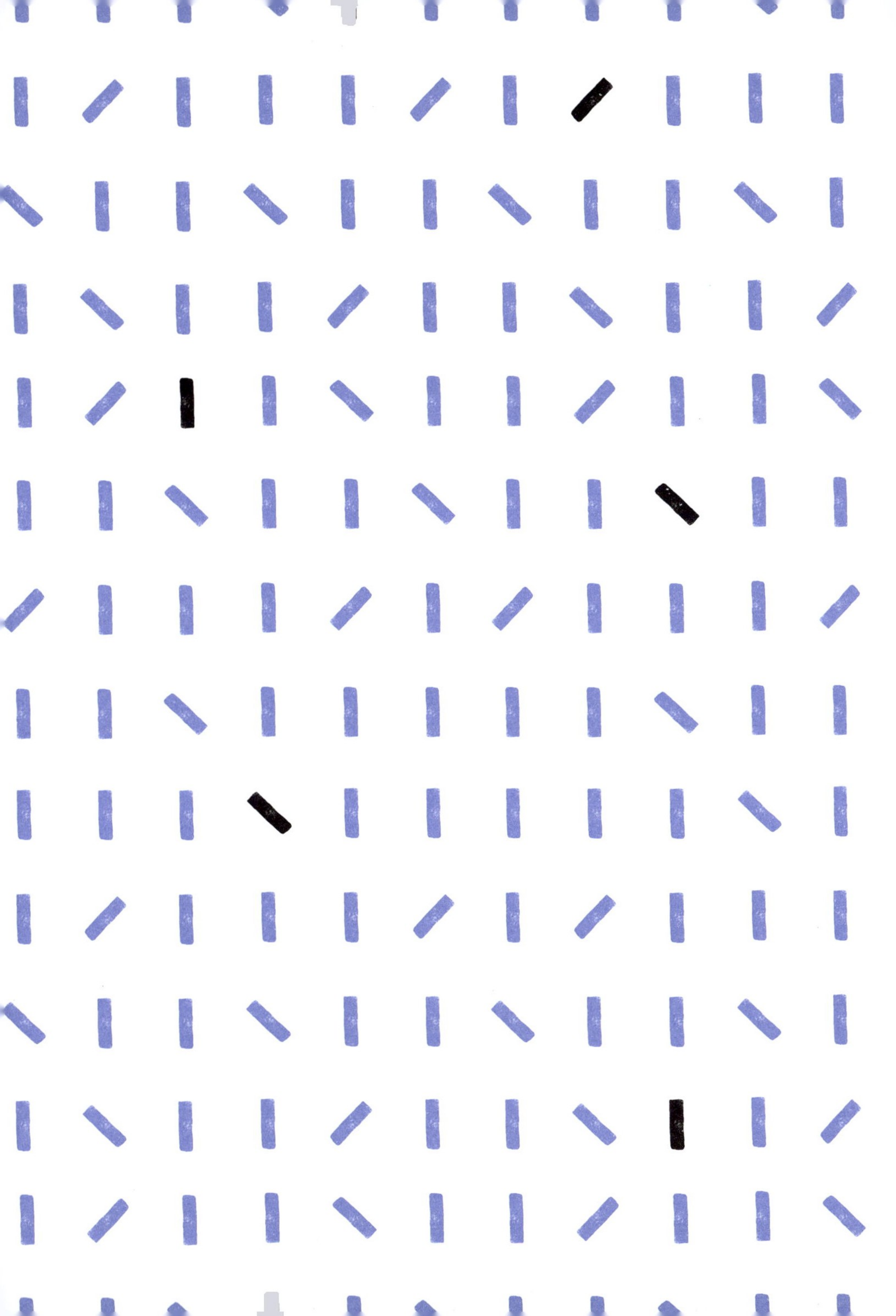